Competing Truths

Competing Truths

THEOLOGY AND SCIENCE AS SIBLING RIVALS

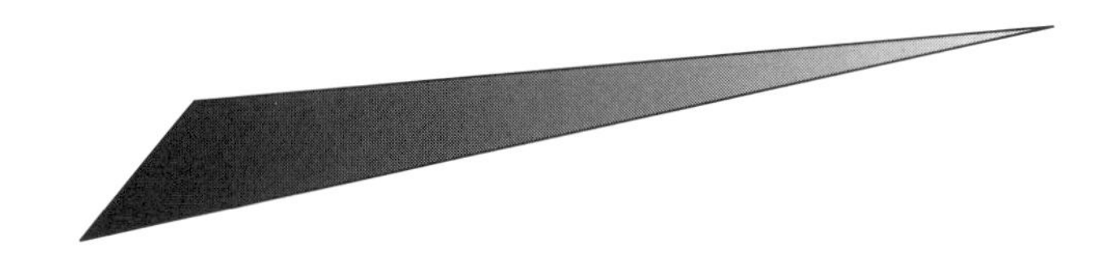

RICHARD J. COLEMAN

TRINITY PRESS INTERNATIONAL

Trinity Press International, P.O. Box 1321, Harrisburg, PA 17105
Trinity Press International is a division of the Morehouse Group.

Cover design: Corey Kent

Library of Congress Cataloging-in-Publication Data

Coleman, Richard J.
Competing truths : theology and science as sibling rivals / Richard J. Coleman.
p. cm.
Includes bibliographical references and index.
ISBN 1-56338-360-8 (alk. paper)
1. Religion and science. 2. Truth – Religious aspects – Christianity. I. Title.

BL240.2 .C643 2001
291.1′75 – dc21

2001027250

Printed in the United States of America

01 02 03 04 05 06 10 9 8 7 6 5 4 3 2 1

Contents

Preface

This book scrapes across the grain of current thought. Both in the popular press and in academic circles, the rapprochement between theology and science continues. The current direction of intellectual flow is toward "methodological convergence" (Nancey Murphy), a "graceful duet" (van Huyssteen), or "hypothetical consonance" (Ted Peters). I am more than just a little wary of where the trend toward consonance, new symmetry, or a coherent system will take us. *Competing Truths* proposes the contrary thesis: that theology and science must be envisioned as not complementary perspectives but rather as rivalrous—indeed, fruitful *because* they are rivalrous.

These pages also contain three core arguments, one historical, one philosophical, and one theological, which constitute Parts One, Two, and Three.

In Part One, I discuss how theology was dethroned by science as queen of the intellectual world. Science, however, did not become the next queen because it failed to answer the perennial questions about who we are and what our place is in the nature of things. The dethroning of science resulted in a rush to fill the vacuum by the "-isms" of the twentieth and twenty-first centuries.

In Part Two, I propose that in the domain of philosophy our postmodern condition is defined by a double crisis in epistemology and ontology. The Old World order of a natural fit between epistemology and ontology is gone. Philosophy broke apart the fit in favor of epistemology, resulting in the turn to the subject. Science moved from a hard to a soft ontology, making it increasingly difficult to believe that things in themselves have intrinsic properties; those properties are what once defined our place in the cosmos.

Part Three discusses theology's current state of being bent out of shape because it has tried to emulate the success of science by adopting an empirical-like methodology. Its recent history is best understood as the pursuit of a valid methodology. In the process, theology surrendered what it has always done best: word-truth.

Between these core arguments, there are other equally provocative contentions. I am not hesitant to talk about sin, word-truth, and the real. All have been dropped from the agenda of theology and science. What if sin is ontologically deep rather than just an epistemological defect? Theology once believed worlds could be created with words and science believed words could be fitted to the way the world is. This is the philosophical foundation that once was and I am not trying to rehabilitate it. On the other hand, I am not prepared to relinquish the unique role that language has to provide meaning and understanding, sometimes apart from the physically real and sometimes because it does reference the givenness of what is other than self.

It is common parlance in the Western world that science displaced theology as the dominant mentality. How that came about is an interesting story. The process of dethroning was not warfare but a rivalry between siblings. For nearly three centuries (sixteenth through the mid-nineteenth), they fought over the same turf. Their engagement—sometimes amicable and sometimes antagonistic—was about three irreconcilable paradigms:

- The metaphysics of perfection versus the principles of mechanics and mathematics,
- Special creation versus natural selection,
- Providence or teleology versus progress.

By the end of the nineteenth century, theology was dethroned, and when all is said and done because science had demonstrated a superior methodology. Science, however, did not become the new sovereign. It was not able to provide a coherent worldview—surely a new view of the world—because it lacked the resources to bring together two incommensurate views of the world: one mechanical and static and the other biological and evolutionary. As a result, science was incapable of answering the perennial questions of what is human nature, what is the nature of the universe, and what is humankind's place or role within the cosmos. The way was open for pseudo-sovereigns to co-opt the methodology of science in order to build a New World order. As science entered the twentieth century, it capitulated from within. There were no rivals to threaten its hegemony, but it was beginning to acknowledge the impossibility of a literal, objective description of the universe. Without the ability to claim a perfect fit between methodology and the way the world actually is, science was compelled to back off its claim to an all-knowing truth. From the outside, science was challenged by historical events, demonstrating that it did not have the answers to all of our important problems. The Challenger and Chernobyl disasters were harsh reminders of what did go wrong in spite of assurances to the contrary. The dethroning of theology and the capitulation of science became the raison d'être for the postmodern chaos of competing ideologies.

The triumph of empiricism is not the entire story. Philosophically, our postmodern condition needs to be understood as a crisis of *both* epistemology and ontology. To be queen requires achieving a fit between how we come to know (epistemology) and the world itself (ontology). In dethroning theology, science eschewed the fit in favor of epistemology—the beginning of our epistemological crisis. Even this is not the full story. The theological and philosophical muddle we now experience has roots as far back as Descartes. Before the Enlightenment, knowledge came "from connecting ourselves rightly to the significance things already have ontically" (Charles Taylor, *Sources of the Self,* 186). Modernity, then, is the transfer of the locus of truth to the interior. What mattered most was the immediacy of clear and distinct ideas, a priori concepts of the mind, and ordering by reason and procedure. Science, however, did not follow philosophy's turn to self and language; it continued to believe there was validity in the model of correspondence—a relationship or fit between words and the world, between theories and what is discovered "out there." The real that was essential for Galileo is still important for most contemporary scientists, except that the real has become relational rather than objective.

However, the crisis is not just epistemological or ontological, but the rupture of the fit itself. Consequently, the predominant model of correspondence between word and the world has been killed by a thousand qualifications and we are left to pursue methodological concerns apart from the meaningfulness of what is real.

What was theology to do? The ontological foundation of a purpose-driven or teleological cosmos had been reduced to "facts." The philosophical foundation of revealed truth known through inspiration and intuition, the epistemology of premodern and modern theology, was also decimated. The authority of word-truth and its power to create a worldview had been pulled into the murky waters of language analysis and the relativity of words referenced to words. Theology was bent like light around a black hole, leading to epistemological confusion and irrelevance.

How does all of this affect the dialogue between theology and science? Let us remember that theology and science have a long history as sibling rivals. Their relationship has been distorted by cliches such as "warfare" and now by the rush to collaborate. Theology was the firstborn and, as the firstborn, the dominant sibling. Science was the upstart brother who labored alongside theology until the time came to surpass her. Philosophy was the middle child, destined to mediate between theology and science, by brokering disputes, clarifying and demanding further clarification. Fundamentally, they have been and still are rivals, because they operate with different epistemic-ontological fits. They also have different governing interests and different domains of expertise.

Competing Truths: Theology and Science as Sibling Rivals has several constructive contributions to make. One is to place the conversation between theology and science in its broadest historical and philosophical context. Another is to encourage theology to bring its tradition of word-truth to bear upon the ontologically real. And, in the paradigm of narrative truth, we have a legitimate way for the siblings to incorporate the word-truth of theological assertions and the fact-like statements of science. Narrative truth has the potential to connect decisive events—wherever they occur—in a way that teases out plot lines. Rather than finding God or not finding God in the causes, scientists and theologians will be better served if they focus on the significance of decisive events across a narrative line.

As siblings, science and theology belong to same family of truth seekers. As rivals they need to challenge each other, for every truth-seeking discipline can only decide what is truth from within the fit that is particular to its own tradition. Collaboration and corroboration are good and worthy; however, they dull the cutting edges that are required to correct myopic vision. My hope is to push scientists and theologians past the paradigms of comparison and contrast, opposition and competition, because the rivalry I speak of is postfoundational, less defensive, more mature. This is not an invitation to raise barriers or to retreat to separate but equal domains. On the immediate horizon are the most critical issues we have ever faced. Pressing ethical questions arising from genetic engineering, cyber technology, and the degradation of our ecology are before us. It is imperative that theology and science share their wisdom and challenge each other in order to shape the future that is ours to shape.

This is the kind of book that could overwhelm the reader with footnotes. I have taken a commonsense approach. Generally, I have cited a direct reference in the body of the book in order to avoid the irksome need to look elsewhere. The footnotes, then, are for additional readings, complicated citations, and further comments. The bibliography at the end of the book allows the reader to find the complete information regarding a book or other citations. It is reserved for the sources that are directly quoted; otherwise, I have cited just the author, the title, and the date when the reference is of a general nature.

This book has been the slow, plodding effort of a parish minister finding time in the midst of other pressing duties. This is not to say there were no supporting hands along the way. I am grateful to my many colleagues who read parts of the manuscript. I am especially grateful to Professor Bill Woodward for his early encouragement, to Dr. Steven Holmes, who helped me reorganize most of the material, to my wife for her trusty red pen, and to my sister for her steadfast faith in me that this book would see the light of day. This book is dedicated to my parents who were not intellectuals but who encouraged me in the Christian faith and who had hearts in the right place.

Richard J. Coleman
rcoleman20@earthlink.net

Part One

The Rivals They Once Were

TRACKING THE RISE of science has been a favorite pastime of philosophers and historians. The value of this account, modest in its justification but not in its proposal, is to understand science and its relationship to theology within the paradigms of sibling rivals and sovereigns. The period of 1500–1860 is more complex than the dethroning of theology by its upstart brother. We must account for theology's attempt at accommodation and the impact of empiricism as the triumph of a new methodology. Historians and philosophers have misjudged the repercussions of that triumph for theology and for our present historical moment.

As queen, theology supplied a worldview. The scientific revolution began as a new view of the world and, traumatic as it was, it progressed by inciting governments and/or dictators to attempt to dominate the world by creating their own worldviews. The postmodern climate is a reflection of a sovereign-less world where even science, although it is the dominant rationality, must make softer truth claims and acknowledge itself as a historically conditioned and limited enterprise.

Chapter 1

Science and Theology as Rivals

Nonetheless, it seems safe to say that modern thought was born in a crisis of authority, and owes its secular character largely to a series of bold attempts to circumvent the dialectical impasse imposed by that crisis.

—Jeffrey Stout, *The Flight from Authority*, 174–75

An Aristotelian theory seems to determine a person's paradigm purposes from the order of nature. To be guided by reason is to be guided by insight into this order. But the modern notion of reason is of a capacity which is procedurally defined. We are rational to the extent that our thinking activity meets certain procedural standards. . . .

—Charles Taylor, *After MacIntyre*, 25

The philosopher is more likely to insist on some semblance of truth; but neither [the philosopher nor the astronomer] will learn or teach anything certain unless it has been divinely revealed to them.

—Anonymous introduction to Copernicus's
"On the Revolutions of Celestial Bodies,"
The Nature of Scientific Discovery,
ed. Owen Gingerich, 301–4

– SECTION 1 –
THE STATURE OF BEING QUEEN

To Be Queen

To be a queen is to rule the intellectual domain and to be respected for the knowledge you possess. Within modern history, there have been only two sovereigns: theology and science. Now, however, there are none. How this came about and what it means to live in an era where neither theology nor science reigns as queen is the subject of Part One. Science and theology were not always rivals. In the extended period of theology's reign, they were collaborators, providing the conceptual framework and the psychological security that made it possible to narrate a coherent story. I speak of them as rivals in the respectful sense that they contended for the same intellectual property, sometimes cooperatively and some times combatively.[1] During the period of science's ascendancy, "rivals" is simply the best way to describe their relationship: at various times they were contentious rivals and at other times they functioned cooperatively. In this section, I will define and introduce terms as well as raise issues that will follow us through the book.

The reader will notice that I nearly always refer to theology rather than to religion as one of the sovereigns. There are several reasons for this conscious decision. The word "religion" is a

---

1. Combatively is as far as I want to push the aggressive nature of their relationship; thereby, I reject, along with most scholars, the paradigm of warfare as misleading and oversimplified. Characteristic of an earlier, somewhat narrow approach to the relationship between religion and science are: J. W. Draper, *History of the Conflict between Religion and Science* (London: King, 1875); A. D. White, *A History of the Warfare of Science with Theology in Christendom* (New York: Appleton, 1896); and J. Y. Simpson, *Landmarks in the Struggle between Science and Religion* (London: Hodder & Stoughton, 1925). Martin Rudwick gives an update on this notion of "warfare" between theology and science; he argues that the traditional stereotype of the conflict model is still dominant in the public mind. "What is striking," he writes, "about this kind of scientific triumphalism is that it remains so persuasive, not only among self-defined top scientists but also among the general public." See Martin Rudwick, "Senses of the Natural World and Senses of God," in *The Sciences and Theology in the Twentieth Century*, ed. Arthur Peacocke (Stocksfield, England: Oriel, 1981), 241–61; in the same volume, S. M. Daecke updates the "conflict" as it is viewed in various ecumenical papers.

much broader and more inclusive term, as in "the religions" of the world. It is so general that it can mean too much or practically nothing at all. After decades of research, Wilfred Cantwell Smith has made a convincing case that the word "religion" is a modern construction. As a construction, it has ossified what are dynamic movements of faith.[2] While we have learned to think in terms of the various religions of the world, there are many traditions that embody numerous cultural expressions of religious experience. By religion, then, I mean the cultural institutions, beliefs, and rituals that form and hold together a community of people, providing them with a distinct identity. In contrast, theology is the discipline whereby we reflect upon the meaning of religious faith. To be even more specific, theology examines the theoretical understandings that underlie religious traditions and compares them with other understandings, whether they are embedded in other religious traditions or, for example, in science. Thus, theology rather than religion is the natural parallel to science, since both are intellectual disciplines with well-defined boundaries.

Theology ascended to the throne by way of a unique synthesis of philosophy, theology, cosmology, and natural science. Without trying to say precisely when theology received the honorary title of "Queen of the Sciences," the kingdom was delineated by the writings of Saint Augustine (354–430) and Saint Thomas Aquinas (1225–74). Under theology's reign, which reached its greatest domination in the medieval period, other disciplines such as law, poetry, philosophy, geometry, music, astronomy, and arithmetic were cast in the role of handmaidens. Even as late as the seventeenth century, in the midst of the Counter-Reformation, the painter Peter Paul Rubens depicted theology leading the way with chalicc in hand followed by four subservient figures: Science, represented as a bearded young man in the background; Philosophy, a bearded old man leaning on a staff; Poetry, looking heavenward; and Nature, of ample Rubenesque bosom.

---

2. Wilfred Cantwell Smith, *The Meaning and End of Religion: A New Approach to the Religious Traditions of Mankind* (New York: Macmillan, 1963).

Figure 1.1. "The Triumph of the Catholic Faith" 1626–28[3]

Trying to date the rise of science is even more difficult, although there is a general agreement that the crucial period was from 1500 to 1860. Contrary to the popular view, theology was able to adapt to a new view of the world (Newtonian) and as a consequence effected an extended period of transition until nearly the end of the nineteenth century. In order to displace theology as queen, science had to forge a comprehensive discipline capable of combining a new way of coming-to-know (epistemology) with a new view of the real (ontology). Science accomplished this late in the nineteenth century, but it is necessary to add a disclaimer. Science became the dominant rationality but it did not create a total worldview comparable to what theology had achieved. That is to say, theology's reign was different; to speak of science as queen could be misleading.

3. For an analysis of this tapestry, see Franklin L. Baumer, *Religion and the Rise of Skepticism* (New York: Harcourt Brace, 1960), 113; and Baumer, *Modern European Thought: Continuity and Changes in Ideas, 1600–1950* (New York: Macmillan, 1977), 63. The picture, Baumer states, "obviously symbolizes theology as queen of all the sciences, the new sciences as well as the old." As the culmination of Renaissance art, Raphael's *Stanza della Segnatura* (a cycle of frescos completed in 1511 in the Vatican

According to Langdon Gilkey, a queen must possess two qualities: first, autonomy from other intellectual disciplines in her theorems, data, and method; and second, the discipline in question must produce the kind of knowledge that is highly prized.[4] In short, a queen provides a worldview which makes it possible for individuals and a community of individuals to be in a changing world in a coherent and consistent manner.

Gilkey's definition is acceptable to a point. I would argue for further restrictions and distinctions. To rule as a sovereign requires that there be a constellation of paradigms coalescing into a worldview (*Weltanschauung*) making possible a new way of being in the world. A view of the world (*Weltbild*), on the other hand, does not provide the same comprehensive fit between the world and how the world is known, although it does present a coherent picture of some dimensions of the world. This distinction is critical when I reject the common notion that at the end of the eighteenth century the Newtonian synthesis, which was a synthesis only encompassing physics, astronomy, and mathematics, constituted a worldview that displaced the Western-Aristotelian-Christianized worldview.[5] A Newtonian view of the world assuredly displaced a Ptolemaic, earth-centered cosmos,

---

apartments) depicts the four cardinal virtues: Law, Theology, Poetry, and Philosophy. In his analysis of the cycle, E. H. Gombrich comments, "all intellectual disciplines somehow partake of the revelation of higher truth. Some indirectly and obscurely, some directly and radiantly. God had spoken implicitly through the mouth of inspired poets and philosophers and openly through the Scriptures and the Traditions of the Church." Here theology is again seated upon a throne, and while the other three virtues are juxtaposed, the tradition in use was not to profane the sacred but to sanctify the profane. See Gombrich, *Symbolic Images: Studies in the Art of the Renaissance* (Cambridge: Phaidon, 1972; New York: E. P. Dutton, 1978), 85–101).

4. Langdon Gilkey, "The Structure of Academic Revolutions," in *The Nature of Scientific Discovery: Symposium Commemorating the 500th Anniversary of the Birth of Nicholaus Copernicus*, ed. Owen Gingerich (Washington, D.C.: Smithsonian Institution, 1975), 538–46.

5. Newton did not, properly speaking, create a synthesis; later generations did so. But Newton was more than an eclectic. He looked for underlying principles in everything he did—a real mark of his genius. From a strictly historical point of view, Newton's ideas were the culmination of the scientific revolution beginning in the 1500s. The rapid acceptance of his scientific principles made a synthesis inevitable. The format of Newton's *Principia*—definitions, axioms, propositions, theorems—was so forceful that other scientists felt compelled to argue from principles; for example, Sir Charles Lyell's *Principles of Geology* (1830–33), and John Stuart Mill's equally influential *System of Logic, Rati-*

but it did not have the capacity to integrate new sciences, especially the biological ones. Because science's reign was principally a triumph of methodology, it created not a new worldview but the possibility for many views of the world, all vying for supremacy.

Worldviews, Paradigms, Epistemes, and Perennial Questions

In this book I will employ several technical terms and a few definitions peculiar to my own interpretation of historical change. Specifically, there is the distinction between a view of the world and a worldview, the technical use of "epistemology" and "ontology," the difference between a paradigm and an episteme, and the role of perennial questions.

The distinction between a view of the world (*Weltbild*) and a worldview (*Weltanschauung*) is a helpful way to understand the complex and larger historical transitions both within science and theology, and between the two as rivals. Nicolaus Copernicus, along with Tycho Brahe, Giordano Bruno, and Johann Kepler, gave us a new view of the universe.[6] Newton significantly extended the work of Kepler and Galileo with principles leading to a new synthesis. Newton wrote his monumental *Principia* (1687) as a new science founded on scientific principles. However, I am reluctant to describe the Newtonian revolution as a true *Weltanschauung*, for a worldview is our most basic way of seeing "what is" and what is possible. We would have an authentic worldview only if this method of knowing (empiricism)

---

ocinative and Inductive, Being a Connected View of the Principles of Evidence and the Methods of Scientific Investigation (1843).

6. Copernicus's place in history has been a subject of much debate. Scholars are inclined to view the Copernican revolution as an invention of eighteenth-century historians. It was not until the seventeenth century, with the additions and corrections of Brahe, Kepler, and Galileo, that the Ptolemaic system had been successfully replaced with something new. See, for example, Bernard Cohen, *Revolution in Science* (Cambridge, Mass.: Belknap Press of the Harvard University Press, 1985), 105–25; Stephen Toulmin and June Goodfield, *The Fabric of the Heavens: The Development of Astronomy and Dynamics* (New York: Harper & Row, 1961), 179.

had succeeded in providing a comprehensive framework that articulated a new understanding of our place in the universe.[7] In addition, it is necessary to take into account the historical movement from the Newtonian synthesis (a revolution on paper) and its translation a generation later into the Industrial Age. The Industrial Revolution certainly constituted a new pragmatism, even a new technological culture; in its secularized form, it displaced the medieval Christian way of being in the world. There was still a crucial component that resisted incorporation into the Newtonian-mechanistic world, and that was the world of chance and evolutionary change.

In future sections, I will frequently describe the historical transitions leading to the dethroning of theology and the capitulation of science as shifts in ontology and epistemology. The usual definition of ontology refers to *that which is,* while epistemology means *how we know* what is. Ontology is the "what" that is discovered outside the human mind. Epistemology, then, is the self-conscious method employed to discover something. What makes epistemology a distinct discipline is the self-conscious reflection required to delineate one method from another.[8]

To be sure, ontology and epistemology usually go hand in hand; they come together as a particular fit. The cosmos was seen as the embodiment of meaningful order that could define the good for us. Theology also operated with a powerful epistemology. A world constituted by ontological essences and ideal types could be known intuitively and by deduction. Theology's epistemology of revelation and inspiration, undergirded by intuition and self-evident truths, was an ideal fit—the cosmos was a harmony of perfect spheres energized by an unmoved mover and each life form had its created place in a Great Chain of Being. A method that has access to the mind of God would know the pur-

---

7. See Charles Taylor, *Sources of the Self: The Making of Modern Identity* (Cambridge, Mass.: Harvard University Press, 1989), 19–24.

8. For the helpful phrase "coming-to-know" and the treatment of epistemology and ontology, I an indebted to Mary Gerhard and Allan Russell, *Metaphoric Process* (Fort Worth: Texas Christian University Press, 1984).

pose or the goal (*telos*) of every thing that existed. If God created the universe for God's purposes, then it made logical sense to believe that the cosmos could be described and explained best with universally valid propositions and theologically correct doctrines.

Likewise, with the ascendancy of science it seemed as if a new perfect fit had been found. An objective world would lend itself to being described objectively; a universe inherently mathematical would be best explained with mathematical precision and mechanical models. As a method of data gathering, induction was the method of choice because truths about the world had to be carefully assembled; a method of experimentation and testing was the appropriate one because empirical facts had to be "teased" out of nature. Science dominated because it had the power to describe empirically what is real, and what is real was known best by the scientific method of empiricism.

This fit, however, does not exclude the possibility that one may take the lead over the other. During its tenure as queen, theology represented the predominance of ontology over epistemology. The world was ordered by God, seeded with ideal types, constituted by goals, and moved providentially as the story of fall and redemption. While theology ruled, individuals found their self-identity by aligning the disorder of their lives with the cosmological order—the way God had created it to be. Science, on the other hand, was foremost the triumph of method and therefore epistemology predominated.

It follows that whether it was the dethroning of theology or the capitulation of science, the underlying dynamic would have to be the dissolution of the marriage of epistemology and ontology by discrediting either epistemology or ontology or both. Galileo, for example, challenged the Roman Catholic Church's sole right to declare what was real or not real, and he did so by advocating a new way of coming-to-know. He only indirectly challenged the Church's right to declare ontological truths. As empiricism demonstrated the ability to predict celestial events and experimental results, the Church had to narrow its truth claims concerning factual information about the world, while science proceeded to expand upon its truth claims concerning the real and the relevant. It was the soft underbelly of theology's method of knowing—soft in comparison to its integrated worldview of providence and special creation—that proved to be vulnerable.

On the other hand, science's capitulation was in no small measure precipitated by theoretical problems in physics concerning the world as it actually is. Principles of uncertainty and relativity meant the universe was not a measurable dimension of absolute particles moving across an absolute grid of time and space. Science also had to scale back its methodological claims and account for the influence of the observer. In the crisis of ontology that confronted theology, its methodology came under hard scrutiny. Science, on the other hand, was forced to revise both its epistemology and ontology almost simultaneously—but it was the Newtonian universe of things-in-themselves that gave the most ground. Nevertheless, science never lost confidence in its ability to know more and more about the physical universe.

Another way to discern the distinction between a view of the world and a worldview is in terms of "paradigms" and "epistemes." Thomas S. Kuhn's book, *The Structure of Scientific Revolutions* (1962), has become the standard source for our thinking about paradigm shifts. In the narrow sense, a paradigm is a set of basic assumptions that guide a research program. In a broader sense, a paradigm is an accepted model of interpretation or understanding. Kuhn prefers to think of a paradigm as more than a theory or leading idea. In his "Postscriptum" of 1969, he defines paradigm as "an entire constellation of beliefs, values, techniques and so on shared by the members of a given community" (2d ed., 175). In other words, a predominant idea or research tradition makes up a paradigm only if it has the capacity to attract and integrate minor trajectories.[9]

---

9. One can speak of a gradation of paradigms, such as macro-, meso-, and micromodels, as Hans Küng does. See Hans Küng and David Tracy, eds., *Paradigm Change in Theology* (New York: Crossroad, 1989), 9–10. My definition is akin to Küng's meso-

A shift in paradigms occurs when anomalies arise and reach such an intensity that practitioners are drawn away from one paradigm in order to adopt a new paradigm that satisfies the anomalies. In the process, old data is seen in a new way and new data is successfully incorporated. Examples of paradigm shifts in science would be the change from Aristotle's analysis of motion to Newton's three laws of motion, or the shift from Newtonian mechanics to quantum physics. In the next chapter, I will depict the dethroning of theology in terms of three paradigm displacements. The three rival paradigms in question are:

Queen Theology	*Queen Science*
Metaphysics of perfection	Principles of mechanics and mathematics
Special creation	Natural selection
Providence or teleology	Progress

In describing several centuries of change, it is necessary to work with a concept larger than paradigms. Another term from Michel Foucault helps clarify the process of dethroning: An "episteme" is a communal definition of what is knowable. It is a historical a priori, which in a given period "defines the conditions of possibility of all knowledge, whether expressed in a theory or silently invested in a practice" (*The Order of Things,* 168). An episteme, then, is both conscious and unconscious and defines the "conditions of possibility" concerning what is true. In contrast, paradigms govern a specific field of science or theology. An episteme involves the idea of science or theology itself. Paradigms guide research; epistemes provide them the epistemology, logic, and assumptions that make their research possible.[10] This distinction adds flesh and skin to Gilkey's bare-bones statement

---

model. His examples of this mid-range use of paradigm are the wave theory of light, Maxwell's electromagnetic theory, and the doctrines of creation and grace. His examples of a macromodel are closer to my use of the term "worldview"—such as the Newtonian synthesis or the Augustinian model of doing theology.

10. See Michel Foucault, *Les mots et les choses: une archeologie des sciences humaines* (Paris: Gallimard, 1966). The English translation is *The Order of Things: An Archaeology of the Human Sciences* (New York: Vintage/Random House, 1994).

that a queen must possess knowledge that is highly prized and autonomous. To the degree that theology and science qualified as sovereigns, and their rivals did not, was a measure of their ability not merely to employ new paradigms but to provide the very conditions for knowing and what could be known. An episteme is no less than a worldview that offers a way of being in the world, predicated on an epistemological-ontological fit.

Foucault's introduction of the term episteme is attractive for another reason. Obviously, Foucault is calling our attention to the importance of epistemology. A paradigm includes a collection of ideas or theories required to describe the world, but it tends to neglect the underlying methodology of knowing itself. A descriptive paradigm takes for granted what an episteme does not. It is my argument throughout this book that we cannot understand the historical movements leading to a new queen unless we understand the critical epistemological changes that accompany them. Science introduced a new method of knowing that "won the day." Its methodology of empiricism became the new episteme; it became the unconscious, accepted historical a priori concerning what is real and how one knows the truth about what is real.

When a community or an entire society eventually begins using a new episteme, it could be assumed that there is a clear choice to be made. "In any given culture and at any given moment," Foucault wrote in *The Order of Things,* "there is always only one episteme that defines the conditions of possibility of all knowledge, whether expressed in a theory or silently invested in a practice."[11] We can say with confidence that during the nineteenth century, two rival epistemes involving different standards

11. It is an interesting historical problem how two worldviews might coexist and what happens when the individual is stretched between two conflicting worldviews. As science began to reign as the dominant way of knowing in the twentieth century, did it provide a singular way of being-in-the-world exclusive of other ways? I think not and therefore I question Foucault's presumption unless we speak of a primary and secondary episteme, both of which are operative at the same time. This is what Kuhn argues when two rival paradigms are in contention, but that is a different argument.

of knowing became sufficiently articulated so that only one could *dominate* as queen. The rivalry was not entirely between science and theology or between scientists and theologians, because there were plenty of religious-minded scientists and deistic-minded theologians on both sides of the controversy. At stake was an ever-clearer choice between two forms of knowing. In *Charles Darwin and the Problem of Creation* (1979), Neal Gillespie describes this choice as between positivism and creationism. Gillespie's acute analysis stops at the end of the 1900s, thus he does not consider the possibility that science did not have the resources to translate a dominant epistemology into a viable worldview. But that will always be the failure of an epistemology that is not fitted together with an appropriate ontology.

Theology grudgingly surrendered its picture of the heavens (a view of the world) by finding ways to accommodate its worldview of special acts of creation and providence to the new sciences of the seventeenth and eighteenth centuries. By the end of the nineteenth century, however, its theological assumptions about the fundamental structure of the universe had become vacuous. Science, on the other hand, had gradually but convincingly posed new questions and new answers that seemed more relevant and practical in a world where divine explanations were unwarranted and not useful. It came to pass, therefore, that individuals, either consciously or unconsciously, would choose either a prescientific or scientific worldview in order to understand their place in society and their position vis-à-vis the natural world.

My historical thesis is that science displaced theology as queen but did not become a reigning queen as theology had been. There are still layers of arguments to be added, but another initial definition is important. A queen exercises autonomy over other disciplines and produces knowledge that is highly prized. A worldview does all of this by successfully answering perennial questions. Historian Franklin L. Baumer provides us with this working definition: "The perennial questions, then, are the deepest questions man can ask about himself and his universe. They are perennial because man cannot help asking them; they are fundamental to his whole cosmic orientation" (*Modern European Thought: Continuity and Change in Ideas, 1600–1950,* 11).

The three questions that I consider fundamental and perennial for the conversation between theology and science are: What kind of universe do we live in? What kind of creature are we? What is our place in the universe?[12]

Charles Taylor puts a more philosophical bent to these questions by speaking about orientation and framework:

> But to be able to answer for oneself is to know where one stands, what one wants to answer. And that is why we naturally tend to talk of our fundamental orientation in terms of who we are. To lose this orientation, or not to have found it, is not to know who one is. And this orientation, once attained, defines where you answer from, hence your identity. (*Sources of the Self: The Making of the Modern Identity,* 29)
>
> Here we connect up with another inescapable feature of human life. I have been arguing that in order to make minimal sense of our lives . . . we need an orientation to the good, which means some sense of qualitative discrimination, of the incomparably higher. . . . But this is to state another basic condition of making sense of ourselves, that we grasp our lives in a *narrative.* (*Sources,* 47)

It is not sufficient, therefore, just to ask these types of questions; they must be asked in such a way that they are connected. To do so is to pursue what I call narrative truth.

Historically speaking, then, we have thrown ourselves into a thicket of issues. Is there anything inherent in the nature of human beings that drives us toward discovering and construct-

12. Don Browning suggested these particular three perennial questions in his forecast of how science and theology will intersect in the future. See Browning, "The Challenge of the Future to the Science-Religion Dialogue," *Zygon* 22 (Twentieth Anniversary Issue, 1987): 35–38. Karl Barth found a way to compress all three questions as one: "Who and what is man within the cosmos?" See Barth, *Church Dogmatics* 3, no. 2 ("The Doctrine of Creation"), 68.

ing narrative truths? Are there moments of self-transcendence or even a transcendent perspective that allows us to orient ourselves to what Taylor describes as "some sense of qualitative discrimination of the incomparably higher"? There is the double-sided question that continues to dog us into the twenty-first century: Could theology continue to address perennial questions without including the real as important? And could science find a way to incorporate the verbal truths of theology and philosophy and preserve the integrity of its methodology?

As different as science and theology are in their primary purpose, they complement each other in that fundamentally they strive to find meaningful ways to frame the world. Intrinsic to the purpose of theology is the quest for what it means to be human; intrinsic to the purpose of science is to define what is real. Insofar as we do not separate these two endeavors, we strive to define our place in the universe by constructing narrative truths about the kind of creature we are and the kind of universe in which we live. To be human, in the final analysis, is to feel at home on the planet Earth situated in this mysterious, ever-expanding universe. This unquenchable push to reach for the stars and to know the truth of our story requires the best that both theology and science can offer.

The difference between how theology ruled and how science dominates is also about the difference between authority and power. Authority is authorized power, while power is unauthorized authority. A queen usually rules because she possesses authority. The president of the United States governs because he occupies an office invested with certain authorities. A queen or the president may also wield power, but this power increases and decreases in relationship to her or his popularity. A challenger usually begins with little authority but with a growing power base. During periods of transition, the power of a rival grows until its paradigms, methodology, or view of the world is more highly prized than that of the reigning queen. Consequently, as the authority of the reigning sovereign deteriorates, the power of the rival becomes the new authority. A deposed queen, such as theology, can continue to wield power because the deposed queen still answers perennial questions. Science was given the keys of authority by popular consensus, but now that same popular consensus is not so sure it wants science to wield unchecked, authorized power. Undoubtedly, the capitulation of science is closely associated with a certain hesitancy and suspicion born of a century of unprecedented violence and misuse of technology.

A Brief Historical Overview

Because later chapters will discuss both historical movements and philosophical ideas, I will first provide a brief overview, starting with some markers for the dethroning of theology and the capitulation of science. I have followed I. Bernard Cohen's fourfold division as a basic outline (*Revolution in Science*, 1985).[13]

From 1500 to 1687 (ending with Newton's *Principia*), observation and experiment were joined together to forge a new scientific method to make revolutionary discoveries in astronomy and mechanics. Experiment, observation, and the mathematization of causality became the basis of our knowledge of the natural world.[14] Science began to challenge the sovereignty of theology with a new way to view the world. Copernicus relocated the planet earth, Columbus changed how one looked at the horizon, and Newton replaced philosophical supposition with mathematical precision.

From 1687 to 1859 (ending with Darwin's *The Origin of Species*), the Newtonian style of science was extended and so-

---

13. I. Bernard Cohen, *Revolution in Science*, especially 77–101. I have followed Cohen in using an upper case "R" when referring to a specific Revolution when it incorporates his four stages of development: the intellectual revolution, commitment to the new method or theory, general acceptance among members of the scientific community, scientists begin to do science in the revolutionary new way. See *Revolution,* chap. 2.

14. Edmund Husserl does not stand alone, but he does have an unusually clear understanding of the mathematization of nature beginning with Galileo. See Husserl, *The Crisis of European Sciences and Transcendental Phenomenology* (Evanston, Ill.: Northwestern University Press, 1970), 23–37.

lidified and new institutions were founded incorporating the new discoveries from the first period. The science of mechanics, astronomy, chemistry, and mathematics were considerably enlarged, but eventually the dominant paradigm became Darwin's principle of evolution, leading to the notion that the way of progress in all sciences is not necessarily mathematical. This set up the unresolved tension between the world-as-machine and the world-as-organism paradigms. During this transition period, theology accommodated itself to a new world picture and science gained respect as the predominant way of knowing what is real.

The third period, from 1859 to 1927 (ending with Heisenberg's indeterminacy principle), spans the great revolutions in physics, chemistry, and the earth sciences. Within this prolific period there was a second revolution in physics associated with the Michelson-Morely experiment, Maxwell, Lorentz, Planck, Einstein, Rutherford, Bohr, Schrödinger, and Heisenberg (known as the Golden Era in physics from 1887 to 1927). This period is characterized by the introduction of "theories and explanations that were based on probability rather than simple causality" (*Revolution in Science,* 96). This period marks the collapse of the mechanistic/deterministic worldview associated with Newton. There were also numerous revolutions in chemistry and genetics.

From 1927 to the present, questions have been raised concerning the fundamental structure of matter, the genetics of life, the origin of the universe, and the very meaning of physical laws. For the first time, and at an unprecedented rate, there is a simultaneous revolution occurring in the physical and the life sciences. The physical sciences are being recast by the successful efforts to formulate and test unified theories. The discovery of the code of life, DNA, revolutionized the life sciences both in theory and in practical applications.

For the most part, I have treated the second period, from 1687 to 1859, as a consolidation of the first and as a transition to the third (1859 to 1927). Working with this simpler outline, the period from 1500 to 1859 marks the dethroning of theology and the crowning of science as the new queen.[15] The capitulation of science begins with the twentieth century, marked by the Golden Age of quantum physics (1925–27) and the dropping of the first atomic bomb (1945).

A few broad observations are worth making here. The first period is often regarded as the Copernican revolution. While Copernicus is often given too much credit, it is true that Copernican astronomy seeded the scientific revolution, which was more truly centered on what Thomas Kuhn calls "the mathematization of Baconian physical science" that occurred between 1800 and 1850 (*The Essential Tension,* 220). This outline reveals how protracted the dethroning of theology was. We cannot consider the process of dethroning to be complete until after Darwin, because only with the publication of *The Origin of Species* was there a clear alternative concerning the mechanism of biological change. The paradigm of evolution fueled the engine of progress while displacing the Great Chain of Being as the central paradigm.

Psychologically speaking, Newton straddled the sixteenth and seventeenth centuries. He epitomized an era which had begun with Copernicus and laid the theoretical foundation for the era to come. The rapid acceptance of his scientific principles made a synthesis inevitable. Not only was a new picture of the world emerging, there was a growing confidence in empiricism as the way of knowing what is real. A comparable revolution in chem-

15. Professionals go hand in hand with institutions. Ian Hacking has proposed that every scientific revolution is concomitant with new kinds of institutions. In part, this is why I have not counted the second period of revolutionary change as a distinct period of conceptual change, but have regarded it as a period of consolidation and extension of the Newtonian style of science and preparation for the revolution in the life sciences. The establishment of scientific academies and scientific specialties within universities served several purposes: to provide a sense of scientific community, to encourage peer review, to keep official records, to encourage professionalism, and to foster the idea that science is a continuing endeavor that requires an accumulation of facts. See Hacking, "Was There a Probabilistic Revolution, 1800–1930?" in *Probability since 1980: Interdisciplinary Studies of Scientific Development,* ed. Michael Heidelberger, Lorenz Kruger, and Rosemarie Rheinwald (Bielefeld: B. Kleine Verlag, 1983), 487–506; and Roger Han, who also stresses institutional change as a major feature of the second revolutionary period in science, *The Anatomy of a Scientific Institution: The Paris Academy of Sciences, 1666–1803* (Berkeley, Calif.: University of California Press, 1971), 275–95.

istry did not occur until Antoine-Laurent Lavoisier and John Dalton's *New System of Chemical Philosophy* (1808); and in geology, until Sir Charles Lyell's *Principles of Geology* (1830–33). It was not until Darwin's empiricism had been attacked and defended that science could claim a universal methodology. The biological sciences were more dependent upon comparison and observation than on experiment and mathematics. There was some confusion about unifying principles. In the meantime, science encountered a universe constituted by indeterminacy and relativity. Theories of physics would have to be modified and so would the discipline's cornerstone claim of objectivity.

Finally, the following review is offered as another perspective on the broader shifts that constitute the period of 1500 to 2000.

The Prescientific Worldview

The grand synthesis of Aristotle, Plato, Aquinas, and Ptolemy made it possible to understand ourselves and our place in the cosmos as part of salvation history, a story told magnificently by Dante and Milton. This synthesis formed a satisfying fit between epistemology and ontology. But what exactly was this fit? Beneath the structure of the cosmology lay a preordained ontology of ideal types or essences to accompany a teleological mechanism. The cosmological order radiated out from the Earth in perfect circles, permanent bodies, an unchanging structure. On Earth, the biological world had its parallel ontology of immutable natural kinds and ideal types ordered according to a Great Chain of Being. Everything in the universe was where it should be, not by chance but by the preordaining decision of God. The Aristotelian view of natural places meant that everything "strove" to attain some predetermined place or purpose. It was a short step for Christian theologians to baptize the Greek notion of a natural propensity toward a natural end with a divine causation leading to a providential purpose. The philosophical thought of Plato and Aristotle converged upon this one point: we need essences, or something like essences (natural kinds), to make sense of the world. It is the place of "man," therefore, to figure out what those essences are; or in the case of the Christian, to discern what God intended.[16]

On the ontological side, there were essences to intuitively know intuitively, so what could be more natural than an epistemology of infallible intuition of immutable essences? The method of deduction, therefore, made sense, since it was possible to begin with first principles and deduce the way the world is. As queen, theology co-opted Aristotle and Plato and added its own epistemology of inspiration and intuition to dovetail with an ontology of special creation. The fit, then, on the philosophical side was between necessary truths knowable by intuition and, from the theological side, revealed truths knowable by inspiration. External evidence was not irrelevant, but it was neither decisive nor essential. The ruling paradigms were the metaphysics of perfection, special creation, and providence.

Roger Hull underlines the importance of teleology when he says, "Galileo and Newton replaced one physical theory with another, but they left the teleological world-picture intact. . . . If teleology permeated Western thought, essentialism was even more important" (*Darwin and His Critics*, 54, 67). Any new paradigm would have little chance of success unless it could dislodge the belief in the biological world with its immutable natural kinds and ideal types ordered according to a Great Chain of Being. It could be said that modernity did not reach the stature of an episteme until both teleology and essentialism had been replaced by alternative paradigms.

In a world that was already perfectly finished and where

16. Daniel Dennett clearly sees the distinction between prescientific and scientific as pivoting upon the acceptance of Darwin's overthrow of essentialism. He writes, for instance, that "even today Darwin's overthrow of essentialism has not been completely assimilated," and if we did we would know that "intentionality doesn't come from on high" but "percolates up from below, from the initially mindless pointless algorithmic processes that gradually acquire meaning and intelligence as they develop." See his *Darwin's Dangerous Idea* (New York: Simon and Schuster, 1995), 39, 205.

everything was perfectly fitted, the only proper response was to contemplate the wonder of it all. In a world where heaven and earth interpenetrated each other, belief in providence and prayer were natural, and narrative truth was taken for granted.

The Modern Worldview

The modern spirit created, reached, invented, and explored.[17] Out of this spirit, and because of it, a new ontology and a new epistemology emerged; there was a new fit.

Above all else, science symbolized a new way of coming to know. It began by making a sharp distinction between a systematic method of discovery and the muddle of magic and speculation. This was the heart and soul of Francis Bacon's *Novum Organum* (1620). This new technique for operating in the world did not introduce reason or eliminate faith, but it did insist on separating matters of faith from matters of fact. Langdon Gilkey sets this demarcation of when religious truths could no longer complete with objective facts at about 1830, when the roots of an evolutionary-driven world were being put down,[18] but it was surely inherent in Bacon's separation of speculation/superstition from empirical reasoning.[19] René Descartes's *Discourse on Method* (in Latin, 1637) described four essentials of

17. See Daniel J. Boorstin, *The Discoverers* (New York: Random House/Vintage Books, 1983); and *The Creators* (New York: Random House/Vintage Books, 1993).

18. Langdon Gilkey, *Religion and the Scientific Future* (New York: Harper & Row, 1970), 14. Gilkey chooses 1830 as a significant date because it marked the publication of Charles Lyell's *Principles of Geology* (actually in three volumes between 1830 and 1833). Lyell's overriding principle or methodology was an unwavering dedication to "uniformitarianism." Stephen Jay Gould has an excellent essay exploring the radiating implications of this principle, especially for the understanding of history. See his *Time's Arrow* (Cambridge, Mass.: Harvard University Press, 1987), chap. 4. Both Gilkey and Gould note that Lyell's principle of uniformity was set against catastrophic geology and its association with a theology of divine intervention. Hence, 1830 is a watershed marker of a paradigm shift and of a decision to be made: either history and the universe are ruled by teleology and divine intervention or by uniform regularity of cause and effect.

19. Perez Zagorin, *Francis Bacon* (Princeton, N.J.: Princeton University Press, 1998); Richard Morris, *Dismantling the Universe: The Nature of Scientific Discovery* (New York: Simon and Schuster, 1983); Morris Berman, *The Reenchantment of the World* (Toronto: Bantam Books, 1981), especially chap. 1 ("The Birth of Modern Scientific Consciousness").

a scientific methodology: (1) breaking down larger problems into smaller ones, (2) arguing from the singular to the general, (3) verifying, (4) accepting only what is clear and distinct. This is similar to Bacon's proposals and is what most people accept as the meaning of empiricism. Bacon represents the move toward objectivity, while Descartes represents the initial thrust in the other direction. For Descartes the certitude of truth had nothing to do with any ontological real. It was located solely in the method itself; that is, in the method as a mental procedure. The ontological side of the fit was eliminated and science became a matter of the correct method. In both cases, the modern self was set free from authoritarian restraints and was permitted to generate its own certitude apart from tradition, Scripture, and the Roman Catholic Church.

Science introduced the craft or technique of intervening, inventing, and creating truth. Intervening meant experimental manipulation and the crafting of apparatus to see behind the "curtain" of ordinary observation. Discovering was carried out in an active voice, because it required assembling, classifying, and organizing, coupled with critically and logically formulating a theory or proposing a law. Further intervening, experimenting, and inventing would verify if the theory or the law was correct. Thus, a spirit of assertive curiosity displaced a tradition of passive contemplation. We were creators and our creations were the proof that we "had it right." We could mold and shape the material world, even at the expense of overreaching, because not reaching at all was the modern equivalent of sinning.[20] Final causes and inherent ends were eliminated. The ultimate goal of knowledge was the complete explanation of all phenomena in terms of their causal connections. There was no longer a hidden veil behind which a creator worked. What had formerly

20. Hans Blumenberg nicely contrasts the changing fortune of curiosity as a vice fanning the fires of pride to a virtue necessary for scientific inquiry. Bacon declared curiosity an essential human right that had to be recovered. See Blumenberg, *The Legitimacy of the Modern Age*, trans. Robert M. Wallace (Cambridge: Massachusetts Institute of Technology Press, 1983), part 2, chap. 6 and chap. 9.

appeared to be the mysterious workings of an unseen hand was actually the mechanical turning of a universe set in motion by the clockmaker-God described by the Deists.

There emerged a new world picture and a new image of our place in it. The universe was best described by mathematical precision. Relentless competition, survival, and random processes of selection replaced a benevolent creation having hierarchies of order. In such a world, the notion that we were the children of God had a rival in the demonstrations that human beings had evolved like all animals. The entire Christian drama, dependent upon the centrality of the earth and "man" being the reason for it all, was reduced to no narrative at all or to a narrative that was flat and devoid of perennial questions. In the place of great heroes like Ulysses and King David are the grim pictures of workhouses (in Charles Dickens's *Oliver Twist*) or the horrors of Western exploitation and inhumanity (in Joseph Conrad's *Heart of Darkness* and *Lord Jim*). What disappears is the sense of tragedy and even fate, because neither the gods nor God are sovereign. Ulysses may have been a tragic hero but at least he was a hero. To the extent that there is belief, it is belief in progress by our own hands, but these hands are stained and sullied.

In this modern drive toward objectivity and interior certitude, all forms of word-truth were demoted. The high regard for rhetoric, the vitality of oral history, the respect for tradition, the value of the sacred written word, were deemed irrelevant and of lesser value than the hard scientific fact or the reasoned argument.

Those who sat in the seats of authority were quickly changing. The king, priest, saint, proprietor, and magistrate were displaced by the political leader, schoolmaster, scientist, capitalist, elected official. Choices between competing moral claims abounded. The very rules for what counted as authority and evidence were being turned upside down. The foundation of theology's reign—first principles, self-evident beliefs, dogmas sanctioned by ecclesiastical authority, the inspired Word of God—grew weaker as new criteria for what counted as reasons for believing something to be true grew stronger; namely, experimental data, evidence of higher probability, demonstration, and principles of verification.[21]

When theology reigned as queen, there was a unity of human knowledge, because theology integrated all the other disciplines. In the transition to the modern era, a dispersion of authority and power took place. Consequently, the sense of community and transcendent footholds were difficult to sustain. In summary the changes that took place were thus:

Premodern (Medieval)	*Modern*
Vertical movement	Horizontal movement
Ordered from above	Ordered from below
Will of God (teleology)	Chance and natural selection
Hierarchy of spheres	Expansion of galaxies
Time as lived	Time as measured (by the clock)
Chain of beings	Evolution of species
Destiny	Choice
Knowledge within limits	Knowledge without limits
Feudal kingdoms	Nation states
Defined roles	Be whoever you want
Natural law	Constitutional rights
Revelation/inspiration	Autonomy of reason/objectivity
Saint, knight, priest, Lord	Hero, lay person, scientist, Corporate executive

Postmodern

The movement from prescientific to modern was an unqualified shift of epistemes—the very foundations of how we come to know and what is known were demolished and rebuilt. The transition from modern to postmodern is not as easily delineated. How peculiar it is that the Newtonian synthesis still determines how an average person views the world, while at the same time this synthesis is as obsolete as a woodstove.[22] Postmodernity is a new

21. Jeffrey Stout, *The Flight from Authority* (Notre Dame, Ind.: University of Notre Dame Press, 1981); Ian Hacking, *The Emergence of Probability* (Cambridge: Cambridge University Press, 1975).

22. A woodstove is symbolic of our postmodern condition because it is an anachronism that still works. In this postmodern era, we hang onto anachronisms because of their simplicity in face of complexity that frequently breaks down.

layer spread out over the old rather than a total displacement. In other words, postmodernity is not yet an episteme or a worldview and yet reverberates with its own Teutonic shifts in ontology and epistemology.

In this brief analysis of the postmodern condition, three characteristics come to the fore: first, there is no worldview; second, postmodern science has emerged; third, there is no longer a fit between epistemology and ontology.

Western philosophy was on a quest for foundations in order to construct a tower of knowledge. For deconstructionists, foundationalism is the enemy to be hunted down as a ruse of metaphysics. J. Wentzel van Huyssteen writes that postmodern thought is not only modern thought coming to an end, it "shows itself precisely in the continual interrogation of foundationalist assumptions and thus always interrupting the discourse of modernity."[23] Van Huyssteen would have us see this interrogation as opening a new common ground for the current conversation between science and theology—an idea not far removed from my assertion that an era without ruling sovereigns constitutes a new playing field.

Postmodernists do not like grand stories or meta-narratives. Modernists may have had some doubts about teleology, but many of them still believed in history as a universally coherent process of development, except for the likes of Proust, Joyce, Woolf, Mann, Faulkner and many others. The modern experiment ends when there is no overarching linear teleology, no dominant direction, no stages of progression, no drama of salvation culminating in a final victory of good over evil. Postmodernists deconstruct everything in order to expose the terrors of what we have become. Michel Foucault writes: "The world of actual history knows only one kingdom, without providence or final cause, where there is only 'the iron hand of necessity shaking the dice-box

---

23. Van Huyssteen, "Is the Postmodernist Always a Postfoundationalist?" *Theology Today* 50 (October 1993): 377.

of chance.' "[24] In place of history we have abyss (Jacques Derrida), or genealogy (Michel Foucault), or fiction (Mark Taylor), or unreason (Michael Gillespie).

The epicenter of the rift between the modern, Newtonian synthesis and a new postmodern science was Niels Bohr's startling declaration that we cannot know, or ever know, the universe as it really is. The aftershocks have not lessened with time, because this was not a statement about the technical limitations of methodology (our instruments of measurement) but about the world's fundamental structure (its ontology). Something drastic had to be admitted; namely, the world could not be known without remainders. The universe is not hard but soft, not deterministic but indeterminate, not absolute but relative. The clearer an electron's position, the fuzzier its momentum. Matter is both wave-like and particle-like. Every measurement influences what we measure. Planck's Constant means that there is a lower limit to what we can know about the very small, and Einstein's understanding of the speed of light serves as an absolute limit concerning what we can know across long distances. Quarks are always hidden, always incorporated into something else. The distortion of gravity in space-time requires a non-Euclidian geometry that means the universe has lost its picturability. There are no "things" or bottom-line substances out there, no simple one-to-one cause and effect happenings, only interrelated, interdependent events changing and evolving in an incurably successive world. In short, the universe is ambiguous and evolutionary and therefore our understanding of it must be interpretive, contextual, and historical. Herein was a major implication for the conversation for theology and science. Philosophically, the event swelled to mean that the only access we have to reality is through language, history, and interpretation, which are in turn shaped by culture, subjective prejudices, and self-interests. Langdon Gilkey concludes: "if both

---

24. Michel Foucault, "Nietzsche, Genealogy, History," in *The Foucault Reader*, ed. Paul Rainbow (New York: Pantheon, 1984), 88–89. The quote within the quote is from Friedrich Nietzsche.

scientific and religious knowing represent symbolic constructions of experience and responses to disclosive encounters with the reality, then it follows that they are, in ways neither one has fully recognized, mutually dependent on each other" (*Nature, Reality, and the Sacred*, 32).

In summary, modern science was predicated upon a hard ontology where there were no remainders and time was reversible, deterministic, and having no local contextual exceptions. The laws of physics were universal: everywhere and at all times the same. As the nineteenth century ended, the modern era had full confidence that science would soon know the truth and the whole truth. In trying to define the heartbeat of modernity, William R. Everdell writes: "smoothness, in fact, was one of the ruling metaphors of the age. Nineteenth-century minds disagreed about almost everything except how much they disliked hard edges" (*The First Moderns,* 9). Here was an ontology of continuity and sameness.

While postmodern science has been well documented, I will summarize its main points here. The trend in science has been from hard to soft. If Newton chose to describe the universe in terms of objects located in time and space, postmodern science speaks of the continual transformation of matter. In fact, Einstein did not like the Indeterminacy Principle because, if true, it would mean reality had an indeterminate edge. The world cannot be described atomistically.

In postmodern science, everything is interrelated. The environmental crisis not only demonstrates that everything is related to everything else—it has required us to live with a sense of limits and to understand that values accompany our empirical decisions. There is a real entanglement not only of "objects" but also of quantum entities and systems.

In postmodern science, the observer and the observed are inescapably intertwined. The Principle of Complementarity was an assault upon the objective observer making objective observations; for example, whether a "physical object" displays wave properties or particle properties depends upon the measurement situation, not on the object itself. In addition, to explain something empirically means to know its past as well as its future, necessitating holistic or narrative explanations.

Postmodern science recognizes both singularity and uniqueness in the universe. Nature, like history, exhibits phenomena that defy the laws of physics and buck the law of entropy. Thus, there are singular events of various kinds—and they sometimes have immense repercussions.

Modern science depended upon reductionism to find causal links. Postmodern science reverses that direction and seeks a holist understanding. Once two quantum entities have interacted, they retain a power of mutual and immediate causal influence upon each other, regardless of how far they subsequently separate. Thus, they constitute a single system known as togetherness-in-separation or the non-locality effect. Holistic complexity is nonlinear.

The insignificant can become the significant. The smallest disturbance may change, in an unpredictable way, a much larger system (for example, "the butterfly effect"—the earth's weather can be affected by an African butterfly stirring the jungle air). Systems are much more sensitive to circumstances than previously expected. Also emerging from the study of complex systems are principles of spontaneous self-organization and pattern making. As a result, science must think of top-down causality as of well bottom-up causality.

On the epistemological side of postmodern science, truth is seen as holistic, like a series of boxes where each one is nestled inside a larger box. One truth needs a larger truth in order to be more completely truth. Truth is also constructed and recontextualized. In his influential book *The First Three Minutes,* Steven Weinberg concluded that we are obliged "to construct our own meanings and purposes," since there is no divine purpose to deduce or given meanings to discover.

In postmodern science, the burden of justification has been reversed. John Locke's understanding of the universe (it takes an intelligence to create intelligence) now bears the burden of proof. In *Darwin's Dangerous Idea,* Daniel C. Dennett assesses the impact of evolution in this way. "Now the challenge to imagination was reversed: given all the telltale signs of the historical process that Darwin uncovered—all the brush marks of the artist, you might say—could anyone imagine how any process *other* than natural selection could have produced all living things?" (*Dangerous Idea,* 47).

Knowledge and information have become the new currency of power. Science forces us to make choices about the application of knowledge at an alarming rate. As a result, our postmodern age is confronted with ethical dilemmas that are vastly more complex, larger in scale, and happening at an accelerated pace.

The unity of truth is displaced by a plurality of authorities. We have lost any sense of the whole and our place in the universe. We are left with a cacophony of competing soft claims and the best one can do is to choose by judging "linguistic performance" (Ormiston and Sassower, *Narrative Experiments*, 134). Just as the world has been "thinned out" of ontological properties, there has also been a thinning down of truth. Truth has become a degree of probability rather than a deductive certainty.

Postmodernity is more than deconstructionism, but we do not want to be deaf to what deconstructionists are trying to tell us. Theirs is an approach meant to loosen our addiction to method where method is knowledge and knowledge is control. A masculine way of knowing, fathered by Greek philosophy, heightened by science, and imitated by theology, it has always been preoccupied with the taming and control of the conscious. In quoting Jacques Lacan—"There is no truth that in passing through awareness, does not lie"—biblical critic Stephen D. Moore sees a connection with original sin.[25] Lacan himself is concerned that

---

25. Stephen D. Moore, *Poststructuralism and the New Testament* (Minneapolis: Fortress, 1994), 77–81.

popular psychology has rendered Freud's revolutionary discovery banal, because the unconscious not only exists, it speaks; it not only speaks, it cannot be silenced. A theologian is more likely to say that sin has never left us. To be just a little Foucaultian, our century is the era of the sensitive barbarian, the civilized sadist, the diplomatic despot.

– SECTION 2 –
THE NATURE OF THE RIVALRY

Introduction

The most common understanding of how theology and science relate to each other is as two combatants. Until fairly recently, historical studies tended to confirm this caricature. Two classic descriptions of this clash are Andrew Dickson White's *A History of the Warfare of Science with Theology in Christendom* (1896) and John William Draper's *History of the Conflict between Religion and Science* (1875). The warfare they depicted was largely centered around two issues: first, whether scientific discoveries could be reconciled with a literal interpretation of Scripture; and second, whether the individual was free to pursue truth without limits. White, for example, tells how churches refused to install lightning rods, because they were viewed as a sign of disbelief in the providence of God. White and Draper depict the warfare as a clash between the authority of the Church and individual scientists who were being recognized as representative of a new intellectual force. In Draper's view, the Christian Church was the culprit, because it repressed scientific thought until science grew sufficiently strong to throw off the yoke. White is more moderate, stressing the harm done when freedom is denied to the scientific enterprise. Draper and White are themselves evidence that the end of the nineteenth century represents a low point in the conversation between theology and science.

By way of contrast, two recent historical studies are reluctant to make generalizations because such generalizations obscure the true complexity and diversity of their interaction. Writing about the Methodist reformer John Wesley, John Hedley Brooke comments that in Wesley one sees how a deeply pious man could be an enthusiast for certain styles of science, and a sharp critic of others (*Science and Religion,* 189). Peter Addinall cautions about recent attempts to understand the critical period of the last half of the nineteenth century as something less than real conflict. The conflict he depicts is not one between winners and losers. Rather, he depicts a gradual surrender by theology.[26] Even contemporary historians, though, have underestimated the way the triumph of science influenced and bent the course of theology. The conflict between science and theology was not a war; it was, however, a rivalry requiring individuals to make choices between competing paradigms.

I think it is helpful to cast the relationship as that of contending rivals where intellectual supremacy was at stake and issues of authority and power were always present. With varying degrees of self-consciousness, theologians and scientists were vying over who would hold the keys to the realm of truth. On a more practical level, the conversation was about contrasting ways to understand yourself and the world around you. The rivalry was at times friendly and at other times adversarial; in general, one or the other characterized extended periods. As might be expected, the point of conflict became most intense when representatives of the Roman Catholic Church acted as defenders of their kingdom. As scientists formed their own separate societies, they, too, had domains to defend.[27] The most notable case is the trial of Galileo Galilei. For the most part, individual participants were

---

26. "There was a real conflict between science and religion throughout the nineteenth century in Britain, and it was passed on by the nineteenth as a legacy to the twentieth century." Peter Addinall, *Philosophy and Biblical Interpretation* (Cambridge: Cambridge University Press, 1991), 4.

27. John Hedley Brooke, *Science and Religion: Some Historical Perspectives* (Cambridge: Cambridge University Press, 1991), especially chap. 5 ("Science and Religion in the Enlightenment").

willing to debate openly their differing opinions, but once they understood themselves as defenders of an institution, they became self-protective.

During the cosmological revolution, when a new picture of humankind's place in the universe was being drawn, there was a dynamic give and take among intellectuals. It was difficult to draw firm lines of demarcation between doing science and doing theology or philosophy. Copernicus, Kepler, Galileo, and Newton did not intend to dethrone theology as queen. Copernicus was personally convinced that his model of the universe was just as divinely infused as the Ptolemaic picture. Kepler felt enthusiastic joy over the thought that his mathematical laws were the visible expressions of the divine plan. Galileo wanted to be understood by the Church as a friendly adversary; he worked diligently to find a compromising position between the authority of Scripture and the power of science to explain and discover. Sir Isaac Newton, as newer studies have revealed, was a deeply religious person who did rational science in daylight but the work of a mystic-alchemist under the cloak of night.[28]

New View of the Universe

The clash between theology and science was first a clash between an established world picture and a new one. What we often neglect to recognize is that while these two views of the universe were incommensurate in many ways, theologians were willing to make accommodations. They quickly discovered that they could retain most of their fundamental theological positions such as special creation and providence; that is, they could still tell the narrative story of divine salvation within a new cosmological

28. It is a surprise to most people to learn that Newton was a closet alchemist. Two excellent books that expose the importance of Newton's religious beliefs and situate him in his historical and social contexts are: Gail Christianson, *In the Presence of the Creator: Isaac Newton and His Times* (New York: Free Press, 1984); and Frank E. Manuel, *A Portrait of Isaac Newton* (Cambridge, Mass.: Belknap Press of Harvard University Press, 1968).

framework. This was true as long as they were considering a world picture. Once that world picture also incorporated a philosophy about the universe, a worldview, then an irreconcilable conflict arose.

During the Middle Ages, a Christian synthesis had been achieved that wedded theology and cosmology. As Max Wildiers observes: "[T]he cosmos had become a Christian cosmos, an integral part of an all-embracing sacred order" (*The Theologian and His Universe,* 80). Copernicus's chief affront was to propose a picture of the world that raised problems for Christian doctrine and for the interpretation of Scripture. How does one explain that Jesus ascended into heaven, the clerics wanted to know, if the earth is revolving around the sun? Giordano Bruno (?1548–1600) had the audacity to dismantle the harmonious heavens, including the imperial abode of God, by scattering the stars randomly throughout an unending universe. What kind of God would create such a nonsystem, and how could a storyteller like a Dante or a Milton tell the drama of salvation without a proper stage? Copernicus was not persecuted. Galileo was put under house arrest. Bruno was run out of Geneva by the Protestants and burned at the stake by the Catholics. Copernicus, however, was not a loudmouth; Galileo did publicize his ideas but recanted; Bruno would neither keep quiet nor recant. When intellectual issues were separated from the questions of authority and their political significance was nullified, it was possible for theologians and naturalists to go about their work. As Newton exemplifies, it was possible to be a Christian and still believe in a radically different-looking universe.

Soon after Newton's death in 1727, cosmology was functionally freed from the grip of an Aristotelian-Thomistic theology. Christians could accept the proposition that a hierarchical system of the heavens was not fundamental to their faith. There was still a God of purpose—as anyone could see in the laws that were written into the universe itself. Newton was not crucified; he was deified. This in itself demonstrates that we were still in a period of transition as the seventeenth century ended. Nor was the eighteenth century decisive. Insofar as the Newtonian synthesis was based upon a mathematical model, it did not change ideas about historical time, because mathematical propositions were timeless and unchanging. As Dobbs and Jacob point out, however, Newtonianism's spread eventually led to a far more secular understanding of time and history. Man was a machine, not a sinner.[29]

There is no clearer demonstration that theology had not yet been dethroned than the staying power of the ubiquitous doctrine of the Great Chain of Being. In its original form, this was Aristotle's ladder of nature, which had been transformed from a zoological sequence of organisms into a religious hierarchy of created beings. Each species was a separate creation placed on the ladder without gap or imperfection. Time flowed in a constant upward direction, revealing the infinite goodness of the great architect. "This conception," write Stephen Toulmin and June Goodfield, "dominated . . . eighteenth-century thought to an extent which it is hard to appreciate today. From Leibniz and Locke, through Addison, Bolingbroke and Pope, Buffon and Diderot, to Kant, Herder, and Schiller; one after another, one finds the most influential eighteenth-century authors accepting this notion unquestioningly" (*The Discovery of Time*, 97). Far from being sworn enemies, science and theology were allies in the all-important belief that design and harmony—and therefore, purpose (teleology)—served to reveal a benevolent creator. There were various way to argue that God disclosed himself in the book of his words and in the book of his works (nature); depending upon their prevailing persuasion, theologians, philosophers, and scientists argued either or both.

Science could find operating room within theology and religious belief could function within science, because what persisted

---

29. Betty Jo Teeter Dobbs and Margaret C. Jacob, *Newton and the Culture of Newtonianism* (Atlantic Highlands, N.J.: Humanities Press International, 1995), 86–87.

was the ability to do natural philosophy or natural theology. What had changed during the intervening years was that Newton understood himself primarily as a philosopher who did science within a theological structure, while William Paley, a century later, felt obligated to marshal scientific evidence to make a theological argument about a worldview that was now in question. Nature or "the natural" was still the common thread in both science and theology, referring to everything from what was observable to a belief in essences and purposes, from linear time to a hard ontology, from a methodology that included induction and deduction to intuition and inspiration. When natural philosophy became suspect as harboring theological notions and unwarranted speculations, it was excluded from scientific realms—along with the meta-questions that had been the bridge between theology and science.

A New Worldview

John C. Greene has a well-formulated thesis: "Thus, by the middle of the nineteenth century all of the materials that were to be forged into a distinctive kind of evolutionary world view lay at hand" (*Science, Ideology, and World View*, 133). Four components coalesced at about the same time to form more than just a picture of the world, for they were utilized to construct a comprehensive frame of reference. In other words, humanity had a new way to orient itself in relationship to the universe; something that a Newtonian synthesis failed to achieve.

The oldest and most general component of this new worldview—elaborated in the seventeenth century by Galileo, Descartes, Boyle, Newton, and others—was the idea of nature as a law-bound system of matter in motion. The second component was the powerful concept of organic evolution driven by the mechanism of natural selection. The principal architects were Herbert Spencer, Charles Darwin, T. H. Huxley, and Alfred Wallace. They arrived at more or less the same viewpoint independently in the late 1850s and early 1860s, although this convergence of opinion soon began to show important differences.

The third ingredient composing this new frame of reference was the British school of political economy being formulated by Adam Smith, Thomas Malthus, and David Ricardo. They championed the idea that free competition in the marketplace, if allowed to operate by its own laws of supply and demand, would produce the wealth of nations and the progress of humankind. Social theorists on the Continent, including Rousseau, Turgot, Condorcet, and Comte, expanded these economic implications to envision nature and history as a single continuum undergoing progressive development. Herbert Spencer broadcast the idea that human progress is the outcome of competitive struggle and synthesized all of these fundamental ideas into an all-embracing philosophy that became known as social Darwinism.

According to Greene, the final component in the new worldview was the system of Lockean epistemology that taught the priority of sense experience as the source of all knowledge. Here was the hope that science, as the new sovereign, could know reality as it really is and use that knowledge to recreate the material world. No one articulated this new worldview with more vigor than Herbert Spencer (1820–1903). Thus, objectivity led to positivism and positivism led to agnosticism (*Science, Ideology, and World View,* chap. 6).

While Greene clearly articulates the proposition that during the latter half of the nineteenth century a new *Weltanschauung* came into force, he misses the driving force behind this worldview. It was not the theory of evolution or any other theory, but the methodology behind them that granted science its position as queen apparent. In *Darwin and His Critics,* David H. Hull points out that Darwin was very self-conscious about the methodology he had employed to arrive at his theory of natural selection and quite sensitive concerning how his fellow scientists would judge him in that regard. Darwin's *On the Origin of Species by Means of Natural Selection, or the Preservation of Favoured Races in the Struggle for Life* (1859) appeared at a time when science in general was debating the proper method of doing science. In 1837, William Whewell published his *History of the Inductive Sciences;* in 1843, John Stuart Mill replied with his influential *System of Logic, Ratiocinative and Inductive, Being a Connected View of the Principles of Evidence and the Methods of Scientific Investigation.* They and others were extending the debate begun by Francis Bacon, and Darwin was quite aware of this. As science ascended, it had to solidify its understanding of empiricism as a universal method of discovering truths about the physical and biological world.[30]

Neal C. Gillespie adds the missing key to Greene's historical analysis.[31] Gillespie argues that a crucial transition was occurring. In order to explain this transition, Gillespie uses the distinction between a paradigm shift and the adoption of a new episteme. During the nineteenth century, the conflict between creationism and positivism grew so contentious that a choice was forced. The positivists believed that the only form of valid knowledge began with the senses extended by the inductive method and verified with additional observations. Creationism, on the other hand, was a mix of intuition, the logic of deduction, metaphysical principles, and a theology of inspiration and revelation. Positivism and creationism were rivals because one explained the world as the operation of purely natural causes while the other explained it as the result of divine action.

The journey from Newton to Darwin tells the story. Newton could still hold together in harmony the Aristotelian paradigm of perfection and the paradigm of a mechanistic universe. Darwin could not. Because he was foremost a philosopher, Newton could be both a mystic and a scientist and blend occult ideas and universal laws. Darwin, on the other hand, found it impossible to

---

30. David Hull, *Darwin and His Critics* (Chicago: University of Chicago Press, 1973) chap. 2 ("The Inductive Method").

31. Neal C. Gillespie, *Charles Darwin and the Problem of Creation* (Chicago: University of Chicago Press, 1979).

be both a Christian theologian and a scientist.[32] Newton was still of an era where theological explanations filled in unexplainable gaps and theological arguments bolstered scientific theories. Darwin was in the vortex of an era, the last half of the nineteenth century, where it was more critical and acceptable to draw clear lines separating the work of theology and science. The rivalry for the status of queen had reached a historical apex where the antithesis between design and chance was too sharp to ignore. Science was rapidly enclosing the physical world with "ironclad" laws and infusing the biological with processes that followed no law at all. Deism and agnosticism were no longer alternatives only for the intellectual, because the incongruity of a benevolent creator and a "red tooth and claw" survival of the fittest were serious food for disbelief and doubt.[33]

Through it all there were many clergy who saw no conflict between their clerical vocation and being a naturalist. It was not uncommon for well-educated clerics to pursue an avocation of natural science. Nevertheless, after Darwin, and even before Darwin as Newtonianism became increasingly materialistic,[34] it became necessary to choose between the two professions. There is no other person who typifies the weight of this strain better than Charles Darwin himself. John Brooke cites a statistic that tells the story: "When the British Association for the Advancement of Science had been founded in the early 1830s, clerics had constituted some thirty percent of its membership. In the period 1831–65, no fewer than forty-one Anglican clergymen had presided over its various sections. Between 1886 and 1900 the number was three" (*Science and Religion*, 50).

We begin to see the origins of an even broader divide: a cultural

---

32. See the acclaimed biography of Charles Darwin by Adrian Desmond and James Moore, *The Life of a Tormented Evolutionist* (New York: Norton, 1991).

33. See James Turner, *Without God, Without Creed*, especially chap. 6 ("The Intellectual Crisis of Belief"). Turner argues that theology made so many accommodations to science that it eventually became impotent.

34. See Dobbs and Jacob, *Newton and the Culture of Newtonianism*, 78–95. The authors understand the coalescence of Newtonianism and materialism to have been especially rapid in France and its association with the industrial applications.

divide between religious liberals and religious conservatives. On the one side, there were liberal theologians like Frederick Temple and liberal scientists like John Herschel; they were no longer interested in defending the infallibility of the Bible or supernatural interventions or even in reconciling two seemingly opposing pictures of the universe. They accepted the right of scientists to establish facts concerning the physical world, but insisted at the same time on theology's right to be the final authority concerning matters of faith and morals. On the other hand, there were religious conservatives who were not willing to concede the separate but legitimate argument. Scripture, they insisted, must be taken at face value as being truthful in all that it asserts or not all. They also would not compromise the belief that revelation and inspiration were superior to empiricism concerning the truths that matter; that is, those truths which undergird one's worldview. Such were the issues at Oxford in 1860 between Bishop Wilberforce and scientist-philosopher Thomas Henry Huxley in their famous debate about human origins.[35] For proponents of creationism, which reappeared during the Scopes trial in 1925 and in recent court cases in Arkansas and Louisiana, not much has changed; now religious conservatives advocate the separate-but-legitimate argument and demand that creationism be taught as an alternative scientific hypothesis.[36]

If the conflict between science and theology is an exchange of crowns concerning distinct realms, it does not require the repudiation of religion as such; it only requires its exclusion as a means of knowing the physical world. This, however, was a distinction with implications not clearly understood at the historical moment, nor today. Positivists thought the important step

---

35. Stephen Jay Gould, *Bully for Brontosaurus* (New York: W. W. Norton, 1991), chap. 26 ("Knight Takes Bishop"). See Brooke, *Science and Religion*, 49–50, for the social implications of this debate.

36. Gould, *Bully*, 416–19, 431, 456; and Michael Ruse, *The Darwinian Revolution* (Chicago: University of Chicago Press, 1979), chap. 13. For a more recent assessment see Robert J. Pennock, *Tower of Babel: The Evidence against the New Creationism* (Cambridge, Mass.: MIT Press, 1999).

was to remove theology from science as a causal explanation but not necessarily to eradicate it from the hearts of believers. After Thomas H. Huxley's *Man's Place in Nature* (1863), it was much easier to separate theology and science into different realms. Those who advocated a narrow view of science as well as those who believed in a narrow view of theology were content to go their own ways. By the beginning of the twentieth century—after several centuries of spirited debate—theology and science became more and more isolated from each other. The first half of the twentieth century intensified the separation, leaving in its backwash the notion that theology and science have always been arch rivals, continually at war and with too many fundamental differences to warrant collaboration.

White and Draper, therefore, were products of their time in that they did not see clearly either the rich dialogue that preceded them nor the forthcoming possibilities. They could not know, and we frequently fail to realize, that just as Darwin was finishing his most significant work, *The Descent of Man* (1871), a new generation of creative geniuses was being born: Sigmund Freud (1856), Max Planck (1858), Alfred North Whitehead (1861), Ernest Rutherford (1871), Bertrand Russell (1872), Alfred Einstein (1879), Arthur Eddington (1882), Niels Bohr (1885). As the nineteenth century came to a close, there was every confidence that the final scientific picture of the universe would be mapped, that the basic building blocks of matter would be discovered, that the last missing links in the tree of evolution would be unearthed, that the course of the electron would be plotted, and that the edge of objectivity would be extended indefinitely.[37] It was this very hope and confidence that motivated the new generation of scientists, yet by the end of their generation each had made a

37. Charles C. Gillispie's *The Edge of Objectivity* (Princeton, N.J.: Princeton University Press, 1960) illustrates the persistent confidence that scientific methodology would progress until objectivity encompassed all scientific endeavors. Historically, just the opposite happened.

contribution demonstrating that the universe is not picturable, predictable, or objectively knowable in the Newtonian sense.

Rival Paradigms

Since the displacement of one sovereign by another is a large-scale change, my use of the term "paradigm" represents something more than a research tradition. A paradigm functions as a dominant unit-idea with the capacity to unite lesser themes into a cohesive force for understanding our place in the world. Arthur Lovejoy coined the phrase "unit-idea," which is satisfactory but not as descriptive as "models of interpretation" (Hans Küng) or "maps of reality" (Heinz Pagels), but even these are not comprehensive enough.[38] At stake in this rivalry—regardless of the synonyms we use for a paradigm change—were issues of truth, relevance, epistemology, and ontology; in short, what truths were worth knowing and how they were to be known.

On the surface, gradual shifts occurred in the unit-ideas that would dominate the intellectual world. At a deeper level, there were the unsettled perennial questions about our place in the universe and the nature of the knower who knows. Each paradigm in itself did not constitute a worldview, although they did function as important views of how the world is. A few decades after Darwin's publication of *The Origin of Species,* these paradigm shifts were generally accepted to the degree that theology had been dethroned by a new worldview complete with a new philosophical foundation.

Again, the three contested paradigms that dominated the conversation between science and theology were:[39] the metaphysics of perfection versus the principles of mathematics and mechanics,

---

38. Arthur O. Lovejoy, *The Great Chain of Being* (Cambridge, Mass.: Harvard University Press, 1936), introduction; Hans Küng and David Tracy, eds., *Paradigm Change in Theology* (New York: Crossroad, 1989), 7; Heinz R. Pagels, *The Dreams of Reason* (New York: Simon & Schuster/Bantam, 1988), 161–62.

39. Hans Blumenberg offers another fruitful set of rival paradigms. They are, in slightly amended form:

special creation versus natural selection, and providence versus progress.

The reign of theology depended on a worldview built upon an Aristotelian metaphysics of perfection, the Christian doctrine of special creation, and a deep and pervasive belief in providence. The first paradigm had a stranglehold on the perennial question concerning what kind of universe we live in. The second served to secure the power of God in the natural world by answering what kind of creature we are. The third paradigm made history and human life intelligible and tolerable by defining our place in the ladder of life. As the upstart rival, science had to demonstrate new paradigms for causality, provide a new explanation for human origins, and supply a new rationale for understanding our place in the cosmos.

The Metaphysics of Perfection versus Principles of Mathematics and Mechanics

Greek philosophy was based upon a metaphysics of perfection. Christianity adopted this philosophy as a manifestation of the perfection of God, whose essence of "being" is immutable. Thus, all creation was a manifestation of that perfection. A metaphysics of being, such as Plato formulated, Aristotle expounded, and Thomas Aquinas Christianized, was adapted by theologians with an enthusiasm and a flexibility hard to imagine. What theologians succeeded in doing was to superimpose the Christian drama of salvation over a metaphysics of perfection. In his great epic *The Divine Comedy,* Dante immortalized both Aristotelian meta-

---

- In the category of teleology as defining the "quality of the world for men"—divine providence versus an endangering quality (or self-preservation).
- In the category of explanation—divine will versus mechanistic activity.
- In the category of eschatology or an account of world history as a whole—salvation history versus secularized paradise or messianic expectations of Marxism.
- In the category of politics—theological absolutism (or divine hierarchy/Great Chain of Being) versus the modern doctrine of the state (or the self-assertion of the individual via contract, consent, liberty, law, and rights).

See Blumenberg, *The Legitimacy of the Modern Age,* xxi–xxv.

physics and the Christian drama. The epic is in the form of a journey that takes the pilgrim from the surface of the earth down into the earth via nine circles of hell, then ascends through nine celestial spheres arriving at last at God's throne. In Western thought there may have been no paradigm more powerful and pervasive than a metaphysics of perfection, because it spawned such maps of reality as an immutable God, an unchanging cosmos, ladders of perfection, ideal types, immutable species, and a priori truths.

According to the Aristotelian philosophy of mechanics, nothing moved by itself. The inanimate things of this world were "moved" by "anima" or soul. Each planet had its own "moving spirit" (*anima matrix*). Thus, every place "exerts a certain influence." For a projectile, it is a perfect arch; for a planet, it is a perfect circle. In the case of inert objects, change is the act of the object realizing its potential as it moves toward its natural place in a world striving to fulfill its purpose. In an orderly and rational universe, there is by definition a God-given place for everything and a reason for everything. A stone falls because it is of a class of heavy things and its place is close to the earth. So too, fire rises because it is of the class of things which are light. Thus, the identity of an element and its locus were bound together.[40]

In the Aristotelian-Christian worldview, it is easy to understand why there could be no missing links, no vacuum, and no unexplained causes. A missing link would be an admission that God had failed; a vacuum, an impossible condition. In a void, there could be neither motion nor place. A stone would have nothing to push against it and would have no place to fall down; a flame would not know to leap up. The goal of philosophy, then, was to describe the right and necessary place for everything. Charles Coulston Gillispie succinctly describes the Aristotelian methodology this way:

---

40. For a discussion of Aristotelian physics see Morris Berman, *The Reenchantment of the World* (New York: Bantam, 1981), 51–55; Berry Casper and Richard Noer, *Revolutions in Physics* (New York: Norton, 1972), chap. 6 ("The Fall of Aristotle").

> So it is throughout Aristotle's physics. It was a serious physics, a consistent and highly elaborated ideation of natural phenomena. It started from experience apprehended by common sense, and moved through definition, classification, and deduction to logical demonstrations. Its instrument was the syllogism rather than the experiment or the equation. Its goal was to achieve a rational explanation of the world by showing how the myriad subordinate means are adapted to the larger end of order. (*The Edge of Objectivity,* 12)

The neutralization of space and time did not come easily or quickly. It came by degrees. The Copernican revolution was greater than the person. Because he accepted most of the metaphysical assumptions of Ptolemy and Aristotle, Copernicus had more in common with the past than with the revolution his name would bear. It is often remarked that in almost every respect Copernicus was an ancient astronomer, not a modern one. Johannes Kepler represented a further de-teleologizing of space. In setting forth the new principle that space is isotropic (that space is everywhere the same) and that matter is inert, Kepler was actualizing the implications of the Copernican idea of an earth in motion. Copernicus and Kepler were still motivated by theological concerns; namely, to construct a more uniform and therefore more *perfect* system than the Ptolemaic one. They also took seriously the importance of the real, for both passed onto their successors a log of astronomical computations that had to be reckoned with when one spoke about the real world.[41]

The paradigm shift from a metaphysics of perfection to principles of mathematics required a philosophical underpinning. The Copernican revolution set in motion a revival of corpuscular philosophy. The whole system of celestial physics made perfect sense, Kepler wrote, "if the word soul (*anima*) is replaced by force (*vis*)." No longer was it necessary to hypothesize that each planet had

41. Arthur Koestler, *The Sleepwalkers* (New York: Macmillan, 1959); Timothy Ferris, *Coming of Age in the Milky Way* (New York: William Morrow, 1988), chap. 4.

its own "moving spirit" or "celestial intelligences (*mens*)." The "little spirits" had been replaced by forces. Descartes was intent upon eliminating all such Aristotelian "occult qualities" of space. He began by asking how a single corpuscle, the most basic element of the universe, would move in a neutral or non-teleological space (a thought experiment par excellence!). This neutral void, anathema to Aristotle, was an immensely important step, because only in such a space was a body free to move in a straight line free from all external forces.

Isaac Newton was able to incorporate all of these advances into a unified theory (*Mathematical Principles of Natural Philosophy* [1687]). His primary explanatory model for motion was a body moving uniformly free of all forces, at a steady speed, across a Euclidian plane. It was a new model of immense influence: matter in motion. From such a theoretical ideal, a mathematical idealization could be formulated. The same inverse square law that governed the attraction between sun and planets worked equally well for a stone falling to Earth. More than any other step toward a new sovereign, this one impressed the public the most because it resulted in accurate predictions. Newton's mathematical solutions left little doubt that a purely empirical approach to the mysteries of the world was not only possible, but more eloquent and believable than faith-based theological explanations.

Thus, a new ontological ground for reality was established—a common universality of laws and principles. By the time of Newton's death in 1727, there were few intellectuals who were not convinced of the truth and usefulness of his mathematical statements. The universe had become a vast, intricate machine held together by a single universal force—gravity; that this force had mathematical precision still astounds. A great reversal had taken place. Instead of explaining the divine handiwork in terms of a philosophical-theological framework, a law-abiding machine would do nicely.

This metaphysics of perfection needed a counterpart in the biological world of living things. The Great Chain of Being and its two corollaries, the sufficiency of reason and the principle of plenitude, became that paradigm. Arthur Lovejoy's classic study of the history of ideas showed how controlling these ideas were—no less so than the paradigms of gravity or the second law of thermodynamics. Lovejoy states that "next to the word 'Nature,' 'the Great Chain of Being' was the sacred phrase of the eighteenth century, playing a part somewhat analogous to that of the blessed word 'evolution' in the late nineteenth" (*The Great Chain of Being,* 184).

At the very heart of the Great Chain of Being paradigm was the principle of plenitude and continuity. God in his overflowing love and mercy created both the cosmic order and living creatures in vast array, but each was created finally and perfectly in such order as to demonstrate an infinitely graded series stretching down to the inanimate and up to the eternal God. To quote Lovejoy, "there are no sudden 'leaps' in nature; infinitely various as things are, they form an absolutely smooth sequence . . . to baffle the craving of our reason for continuity everywhere" (*Great Chain of Being,* 327). The principle of sufficient reason, which was often declared the first and basic truth of metaphysics, required the actualization of the ideal in the created world insofar as it was possible. The principle rested upon an implicit faith that the universe is rational order and that nothing happens arbitrarily or by chance. Taken together, the principles of plenitude and sufficient reason presuppose a world where "everything is so rigorously tied up with the existence of the necessarily existent Being, and that Being, in turn, is so rigorously implicative of the existence of everything else, that the whole admits of no conceivable additions or omissions or alterations" (*Great Chain of Being,* 328).

The educated person of the eighteenth century recognized that the ape stood next to man on the ladder of nature, but this was not threatening because "rational life" had been created separately and humankind was distinctly superior to the apes. Just as no organism could develop into a more complex form, no organism could become extinct, since each had been perfectly fixed in its proper slot from the beginning. The very idea of evolution would face no greater obstacle than overcoming a worldview so static that phylogenetic relationships between different species were simply inconceivable until the whole system was dismantled.

Since perfection had to be actualized at every level and in every detail, the idea of purpose, then, is never far removed from the principles of plenitude and sufficient reason. In the Platonic tradition, the idea of "essences" captured this sense of perfection. Teleological purpose was a static conception of progress, because time was frozen into particular acts of creation and providence. Every value was ontological in nature. Christians could scarcely escape being idealists and essentialists, because the necessity of natural kinds, intuitive ideas, and essential qualities was indisputable. Perfection also left nothing more to be explained. For a growing number of scientists, it was just another ad hoc explanation that effectively killed empirical inquiry.

Special Creation versus Natural Selection

Special creation was the most visible focal point of providence, just as natural selection was the heart and soul of Darwin's understanding of evolution.[42] The corollary to the general doctrine of providence (all things work for the good of their creator) was the doctrine of special creation (the creator has made all things good). Special creation was a particularly Christian doctrine in the parallel sense that natural selection was a specific scientific theory; that is, they were paradigms that provided justification for the larger epistemes of providence and evolution. To the extent that special creation became discredited and natural selection "won the day," the Christian worldview of design and designer became suspect and the scientific worldview of life prevailed.

In *Charles Darwin and the Problem of Creation,* Neal C. Gillespie states: "The core of special creation, then, was not miracles

---

42. Neal Gillespie distinguished between four meanings for special creation. See his *Charles Darwin and the Problem of Creation,* 22–25, 147–52.

(although individual naturalists did appeal to them), but the direct, volitional, and purposeful intervention of God in the course of nature, by whatever means, to create something new" (22). The theological doctrine of special creation permeated natural history so universally that naturalists were often unaware of the extent of its influence. There were several reasons for this. Aristotle's influence in biology lasted well past the Copernican revolution in that his immanent teleology was the best explanation for why the living world seemed to strive toward purposeful ends. It was logical as well as comforting to believe that God made each species, that each species was a perfect and marvelous archetype, and that each species was properly placed in the Great Chain of Being. Even into the first half of the eighteenth century, there were no viable rival explanations for the origin of species. Without a substantial fossil record and a hypothesis of natural selection supported by a mass of observational evidence, the concept of special creation explained the unexplainable.

In the latter half of the nineteenth century, a majority of philosophers and scientists abandoned the notion of special creation. Still, they were reluctant to jettison the broader belief in providence as David Hull points out: "Whitehead, Dewey, and Peirce all espoused creative evolution. Alfred Russel Wallace, Asa Gray, and Charles Lyell belabored Darwin with arguments for admitting divine providence, especially with respect to man's mental and moral faculties" (*Darwin and His Critics*, 64–65). Peter Bowler makes a strong case that the "Darwinian revolution" turned out to be a non-revolution. In other words, evolution as a concept could function better without the theory of natural selection and most scientists preferred it that way because it allowed for the idea of progress without the distasteful and mysterious operation of natural selection. Bowler further explains that evolution as a scientific theory incorporating the biological explanation of natural selection did not become established until the 1930s, when a new theory of heredity was available (*The Non-Darwinian Revolution*, 105).

In the scientific explanation for the origin of species, the miraculous intervention of God suffered a fatal, but not necessarily quick, blow at the hands of Charles Darwin. The crux of the matter then and now—and this is what makes it a perennial question—is whether God directs and sustains directly or indirectly. When physical laws were introduced as valid explanations for the transformation of species, the argument of special creation was adapted to include divine action in continuity with divine law. The introduction of the doctrine of catastrophic geology in the middle- to late-eighteenth century illustrates the difficulties in dislodging special creation. Essentially, catastrophism interpreted the stratigraphic features of the earth as successions of sudden, violent disturbances interspersed with periods of calm. At the close of each catastrophic upheaval, life was supposed to have begun anew and continued until the next upheaval. The theory received considerable attention for a short time, because it preserved the assumption of special creation as a series of similar dramatic interventions taking place separately but successively at different geological times. Depending upon your propensity toward supernaturalism, those "special creations" were either supernatural or simply natural in origin. In his *Theorie de la Terre* (1749), Buffon interjected a progressive dimension into change itself. Although his picture looked more like a staircase than a gradually developing tree, he paved the way for a theory of interrelated progressions from the simple to the more complex.

Many scientists accepted Darwin's evolutionism; that is, they accepted nature as moving forward in incremental steps, requiring a dramatic shift in mindset from a paradigm of being to one of becoming. "The nineteenth century," writes Baumer, "may therefore be called the first real Century of Becoming" (*Modern European Thought,* 264). This was a conceptual revolution with serious implications for any theological tradition wedded to an ontology of being. Increasingly humankind was seen as part of nature—not as an exception to the rule but as its epitome. By 1929, Alfred Whitehead was reluctant to envision God as an exception to becoming. His famous principle was: "God is not to be treated as an exception to all metaphysical principles, invoked to save their collapse. He is their chief exemplification" (*Process and Reality,* 521).

The paradigm of special creation also had its historical dimension, since it functioned to explain the Judeo-Christian drama of salvation. The issues were the same. Special creation pointed to the direct, volitional, and purposeful intervention of God in the course of human history, by whatever means, to reveal the divine intention for humankind and for the world. Miracles were the historical events that gave revelation visibility. It was becoming equally difficult to defend revelation—either in the book of Scripture or in the book of nature—because both were understood as the outcome of historical processes that did not warrant a supernatural explanation.

Because it was paramount to rejecting the doctrine of revelation, theology strongly resisted relinquishing the paradigm of special creation. Embedded within special creation was both a theology of revelation as well as a theology of the nature of things. Acts of special creation were spotlights pointing directly to the divine hand. Those same acts were the visible events of redemption—without them, biblical theology scarcely had a story to tell. Special creation also provided a commonsense explanation for different species and put human beings at the top of the Great Chain of Being. In contrast, natural selection flattened out biological processes because change took place so slowly and minutely that it was imperceptible. Where was the designing influence of God to be found? At best, it was hidden in the universal laws of physics and biology, but it was difficult to observe revelation when everything else in the universe developed "naturally." When natural selection was eventually accepted as the driving force of evolution, a place for chance had to be found. Darwin was accused by his critics of breaking the link between the moral and the material world. Without design, a material science was almost inevitable and theology would need to chart its own course. Providence had to compete with progress as a motivating paradigm, and providence was unable to incorporate chance as a fundamental law of the universe.

Providence versus Progress

As one of the most enduring paradigms of human history, providence was being discredited while another was being acclaimed. Anyone who thinks that there was merely a fine distinction between providence and progress—after all, both provided a rationale for human life—misses one of the most pivotal shifts of the nineteenth century.

Providence and progress were competing paradigms. Fundamental to the discussion were radically different conceptions of time. Providence was pictured as a series of still pictures depicting divine events of revelation that, when taken together, told the story of God's redemption of humankind. Stained glass windows reflected a belief that the unchanging and eternal framework of time was interrupted by episodic or inspired acts of God. In contrast, progress was seen as a linear upward movement. Time was continuous and uniform, advancing by degrees and small steps.

Providence was a God-given destiny. Progress, on the other hand, was what humans created. Belief in providence did not exactly foster passivity but, at best, life was a journey through various obstacles to an appointed goal. John Bunyan's widely read and influential *Pilgrim's Progress* (1678) depicted a pilgrimage from the City of Destruction to the Celestial City. What made the journey meaningful was the spiritual destiny at the end. To live heroically along the way required such virtues as obedience, faithfulness, discipline, and honor. It was not that human life was insignificant but individuals endured time, heroically and faithfully; they did not make their own providence, or change their fate in the Greek sense. The Christian was called upon to live up to providence, that is, to seek the spiritual destiny God had laid before him or her. Between the Middle Ages and the Enlightenment, however, complacency gave way to discontent, discontent to protest, and finally, a certain optimism appeared. When people began to achieve change and progress by reform or by revolution, it was an indication that "time was on their side." The necessary virtues became hard work, an inquiring and open mind, and critical thinking. Overreaching was not the sin; not reaching at all was.[43]

According to John B. Bury, original sin stood guard at the door of a gradual and indefinite process of development and moral perfection (*The Idea of Progress*). It was not until evil was discounted and space-time neutralized that visions of utopia could be translated into grandiose schemes for the perfection of the human species. Mathematical principles underlying nature were outside the reach of human sin. Since humankind was superior to the environment, so the logic went, they could build a better tomorrow by reshaping their environment. With each passing decade, science was handing over to humankind not only a natural world subject to mathematical laws, but also a world that could be "vexed" until it yielded fruit of every size and description.[44] Carl Becker sharpens the contrast between providence and progress in this way:

> But if the celestial heaven was to be dismantled in order to be rebuilt on earth, it seemed that the salvation of mankind must be attained, not by some outside, miraculous, catastrophic agency (God or the philosopher-king), but by man himself, by the progressive improvement made by the efforts of successive generations of men. (*The Heavenly City of the Eighteenth-Century Philosophers*, 129)

For progress to displace providence, time would have to be historicized. As Stephen Toulmin and June Goodfield note, the

43. Compare Blumenberg, *The Legitimacy of the Modern Age*, chap. 6 and chap. 9.

44. "Vexing" nature is the way Bacon understood the work of science. It was anything but passive. It was not God who should intervene, but man. Morris Berman writes about Bacon: "Vex nature, disturb it, alter it, anything—but do not leave it alone. Then, and only then, will you know it." See Berman, *The Reenchantment of the World*, 17.

nineteenth century was in every way the "Century of History." The establishment of a more historical conception of the world was only the first step. The second was more difficult—"that of recognizing that the actual course of history was too complex to be fitted into any simple teleological scheme" (*The Discovery of Time,* 233). Both a literal reading of Scripture and Scripture's timeless quality took some hard hits as time was rediscovered as deep and deadly. Loren Eiseley notes the reluctance to accept what now seems commonplace: the extinction of a species.[45] Historical criticism was possible only because time was successive: just as there were geological layers to read, so were there historical layers of interpretation to uncover and compare. The Bible must be read like any text: historically and as literature. In addition, the nineteenth century uncovered serious questions for the next century to answer: Where to locate the historical Jesus? How to understand inspiration? How to assert both the uniqueness of the Christian revelation and still have it meet the criteria of rational (scientific) minds?[46]

The final battle lines between providence and progress were drawn in the life sciences, especially during the critical period of the first half of the nineteenth century. Along with a general belief in providence was faith in special providence: nature itself was described as a miracle and historical events as instances of salvation. Michael Ruse chronicles how quickly belief in miracles fell into disrepute during the several decades before Darwin and Wallace published their theories of evolution (*The Darwinian Revolution,* 87–92). However, one can see the pillars of special providence being torn down even in Newton's lifetime. The neutralization of space and time and its mathematical formalization had the effect of severely limiting the need for divine intervention. Kepler, Galileo, and Newton had moments when they called upon God

45. Loren Eiseley, *The Firmament of Time* (New York: Atheneum, 1984), 36. The reason for the reluctance? "The benignity of Providence," Eiseley replies.

46. By the twentieth century, there were even more ways to read the Bible: culturally, anthropologically, archaeologically, and mythologically. See David H. Kelsey, *The Use of Scripture in Recent Theology* (Philadelphia: Fortress Press, 1975).

to tinker with creation, but this was happening less and in less significant ways. Newton's God had little to do with being the benevolent father who intervened on the behalf of the sick and the needy. The Trinity seems to have been especially problematic for him. This God of the universe had little or nothing to do with living matter. It was almost as if the frailty of human life was covered over with the vastness of the universal plan. Consequently, the suffering and misery of daily life was minimized and simply not in need of theological explanation.

Conclusion

As queen, theology was difficult to dethrone if for no other reason than that it had provided meaningful arguments for existence. Whether Augustine's *City of God,* Thomas Aquinas's *Summa Theologica,* or the Great Chain of Being, the persistent message proclaimed was a divine plan written into the very structure of the universe. Ernest Becker gives one of the better descriptions of theology's worldview that made it possible for humankind to locate itself. It functioned as a worldview because it answered the perennial questions about the kind of world we live in, the kind of creatures we are, and our place in the universe.[47]

> When man lived securely under the canopy of the Judeo-Christian world picture he was part of a great whole; to put it in our terms, his cosmic heroism was completely mapped out, it was unmistakable. He came from the invisible world into the visible one by the act of God, did his duty to God by living out his life with dignity and faith, marrying as duty, procreating as a duty, offering his whole life—as Christ did—to the Father. In turn he was justified by the Father and rewarded with eternal life in the invisible dimension. Little

47. By designating these three perennial questions, I am not excluding others; these do seem especially pertinent to the conversation between theology and science.

> did it matter that the earth was a vale of tears, of horrid sufferings, of incommensurateness, of tortuous and humiliating daily pettiness, of sickness and death, a place where man felt he did not belong. . . . Little did it matter, because it served God and so would serve the servant of God. In a word, man's cosmic heroism was assured, even if he was nothing. This is the most remarkable achievement of the Christian world picture. . . . Christianity took creature consciousness—the thing man most wanted to deny—and made it the very condition for his cosmic heroism. (*The Denial of Death,* 159–60)

Because teleology had been part of the conceptual framework of the Western world from ancient Greece until the time of Darwin, it did not die with the demise of providence or special creation. Instead of dying, teleology was secularized and quickly shorn of its religious roots. Philosophers such as Hume had destroyed the logic behind the design argument well before Paley wrote his popular *Natural Theology* (1802). It was not so much what scientists believed but how they actually practiced science that eventually killed the "scientific" form of the design argument. Darwin's *Origin* was an exemplar of this change in science. For a positivist science in which all natural processes were to be understood in terms of law and natural causes, a theology of providence, which invoked a divine engineer in any capacity, was a causal redundancy. As the argument from design faded, it was resurrected in the form of improvements built into the evolutionary process. Even a minimalist position that nature only produces better-adapted species implies an improvement in that species. A deeper reading, however, reveals that much more was being advocated, a paradox never fully explained. While nature is never said to be aimed at specific goals, it still moves toward "higher" or "more advanced states." The uniqueness of human life, the evidence for more advanced civilizations, the cumulative character of scientific knowledge—all were phenomena in need of an explanation. Even if scientists could not explain it, just as Newton could not explain how gravity worked, they were reluctant to deny the "fact" of progress. Julian Huxley, like his grandfather, was not the only one compelled to live with "the paradox that nature, though devoid of aim and purpose, yet moves toward ever higher levels of order and value" (quoted in Greene, *Science, Ideology and World View,* 165).

Hans Blumenberg points out the final irony. In a world that was already finished, there was no room for human action; one can either adapt oneself to this completed ontology or violate it. Blumenberg argues that Kant, with his great cosmological speculation of 1755 ("Theory of the Heavens"), was the first to propose the idea of an "unfinished world" and to project the archetype of endless progress (*The Legitimacy of the Modern Age,* 212). The idea was certainly inherent in the principle of plenitude and had a firm grounding in the theology espoused by Kant that it would be "'absurd to represent the Deity as passing into action with an infinitely small part of His potency... as shut up eternally in a state of not being experienced'" (cited in *The Legitimacy of the Modern Age,* 212–13). The disappearance of inherent purposes, the demoting of ontology in favor of epistemology, marked the shift from a contemplative role in a teleologically determined world to a constructive role in a world still needing completion. Frank B. Farrell describes the change as a contract whereby "we would give up trying to understand the world in its possible metaphysical depth and would gain greater power to alter and control its appearances" (*Subjectivity, Realism and Postmodernism,* 156). Ironically, the medieval individual was created a "little less than the angels," but at the same time reduced to a rung on a hierarchical ladder; unless you were a king, you suffered passively through life. In the Industrial Revolution, your position was changed from a stationary rung to a moving cog. Now the work of your hands mattered, but you were nonetheless a cog in a great impersonal cosmic clock, laboring under the tyranny of time in an impersonal factory.

Along with curiosity, self-assertion, and critical reasoning, the drive to create a new world was turned over to any who believed that armed with the power of technology they could mold a world better than nature itself. Nature, in its waste, inefficiency, and inhumanity, could be improved upon.

– SECTION 3 –
THE DETHRONING OF THEOLOGY AS QUEEN: THE TRIUMPH OF METHOD

In brief, the dethroning of theology as queen of all knowledge went like this: First, the rival tradition of science emerged and demonstrated a superior ability to ask and answer radically different kinds of questions about the physical world. Second, the static paradigms of a theological-philosophical synthesis were superseded by the dynamic and law-abiding principles of the new science. Third, the teleological arguments from design were discredited by an understanding of a world generated by processes that are self-contained and random. Fourth, eventually a new episteme emerged, which left theology's authority greatly reduced and its relevance suspect.

Among the many things the dethroning of theology signified, the most important was the reversal of ontology and epistemology. The triumph of method was a titanic shift whereby epistemological claims concerning how we come to know took priority and dominated ontological claims about what is true. This book calls theology and science back to an ontological grounding. If this call is going to be heard at all, our preoccupation with methodological questions and the thinness of our ontological claims among all three siblings must be acknowledged.

The scale had tipped in favor of epistemology, but science continued to pursue an objective description of an objective world (the model of correspondence). It is this very same model of correspondence that has been severely criticized and rejected by various scholars in contemporary philosophy.

A New Methodology Is Generated

The discussion of how to define the scientific method has never been settled. There were, however, pivotal individuals who began to see clearly how to distinguish between two contrasting methodologies.

For Francis Bacon (1561–1626) a vast, undiscovered world could be apprehended and put to use—if one could move past the Scholastic belief that reality was revealed in Scripture and extended by deductive principles. The door to this new world of knowledge would be opened by a new method of discovery and reasoning, a *Novum Organum* or "New Instrument" (1620). The new method of reasoning would consist of displacing theology's highly prized method of deduction with an inductive method. Reasoning would be reversed so that certainty would proceed from the ground up by the careful gathering of data free of metaphysical assumptions. In addition, Bacon proposed a new method of discovery. Bacon was convinced that keen observation was insufficient. He writes, "the secrets of nature reveal themselves more readily under the vexation of art [technology] than when they go on their way" (quoted from Berman, *The Reenchantment of the World*, 17). Here was a patent invitation to raise technology over philosophy and theology, experimentation over logical deduction.

This new way "to get at truth" began with an intense, previously unseen skepticism directed at the Aristotelian-Scholastic metaphysics of perfection.[48] At work was a new assumption that hidden behind each occurrence were natural causes, not essences.

---

48. Bacon had no use for the Aristotelian tradition because Aristotle "tried to provide verbal solutions to every problem without attaining any real knowledge." If Aristotle had corrupted reason with dialectics, then Plato had corrupted it with ideal forms. See Perez Zagorin, *Francis Bacon*, 34. With Bacon's sweeping indictment of the past came a negative attitude toward word-truth that only increased until it became incarcerated within modern philosophy.

The scientist would accumulate facts, conduct crucial experiments, and reason inductively. In Bacon's opinion, the theological method of knowing was only a hindrance, since it was only capable of producing speculations and was weighed down with generalities, superficial observations, and fables about the world. It was better to throw off the yoke of philosophy and theology and separate the new science from religion.

René Descartes (1596–1650) was no less adamant about establishing a new foundation for human knowledge. Descartes did not so much offer a system as a style of thinking. He believed passionately in the power of a disengaged rationality. His models were those of mathematics and geometry characterized by rational demonstration, strict methodology, clear and distinct ideas. It was necessary, Descartes concluded, to discredit medieval theology because radical doubt precluded reasoning from the certainty of God to the certainty of the world. This was wrongheaded, Descartes continued, since knowledge of the real world is to be trusted only after it has been conceptualized by the mind and then only after the conceptions have been purified of ad hoc assumptions. A Cartesian distinction would have to be made between natural truths and revealed truths.

Anders Nygren traces the importance of this distinction. Revealed truth is rooted in the rationalist tradition where truths of reason were considered to be necessary and eternal because they were discerned by reason alone (intuitive certainty) or by faith (inspiration). They are called rational, Nygren writes, because they are "true in all circumstances and can never—never in all eternity—be anything else."[49] Natural or empirical truths represent the other fundamental Western tradition where truths have an on-

49. Anders Nygren, *Meaning and Method: Prolegomena to a Scientific Philosophy of Religion and a Scientific Theology*, trans. Philip S. Watson (Philadelphia: Fortress, 1972), 78–102. Nygren writes: "It was in fact a disastrous step that was taken when Leibniz enunciated his twin concepts of *vérités de raison* and *vérités de fait* and the theory of different sorts of truth gained acceptance in philosophy" (82). According to Nygren, this distinction did not originate with Leibniz; it was in the air, so to speak, but was characteristic of his thought.

tological anchor in the way things are as observed or experienced. Nygren rightly decries the philosophical mess that resulted when one kind of truth was reduced to the other and when one kind of truth was judged by the standard of the other. Nygren, however, does not appreciate the enormity of the reversal that was taking place as these two rivals contended. Rational or revealed truths that were once considered eternal were being regarded as contingent, while empirical truths, once considered contingent, were understood as timeless. To add further insult only empirical truths were accorded the status of being reasonable. From hindsight we can understand how the very idea of different kinds of truths invited one being pitted against the other and each having its defenders. "Truths" in the plural remain nevertheless an open and troubling question.[50] Here began the harmful convention that only empirical truths are reasonable. Empirical truths depended on observation and the autonomous self, while truths of faith, or word-truth, depended on the inspiration of the Holy Spirit and on the authority of tradition. In the philosophical tradition flowing from Leibniz, Descartes, and Kant, however, truths of reason were not demoted. They were valid albeit only after they had been purified of subjectivity and grounded in observation.

Galileo Galilei (1546–1642) dedicated his life to defining a new science of knowing. Morris Berman's study of the seventeenth century leads him to assert that the birth of a modern scientific consciousness began with the "highly ingenious combination of rationalism and empiricism which was Galileo's trademark" (*The Reenchantment of the World,* 27). Beginning with a fundamental presumption that the basic laws of nature must be mathematical, Galileo expected a harmony between mathematical forms of knowledge and verifying experiments. Thus, Galileo devised experiments, such as his famous inclined plane, to measure distance

---

50. This issue is by no means dead. It remains alive for those who continue to regard revealed truths as eternal and scientific facts as contingent. It is an issue that divides conservative and liberal Christians, as well as fundamentalists and liberals in many other religious traditions.

against time. He also added another new ingredient—the introduction of scientific abstraction (for example, frictionless planes, pulleys without mass, and freefalls with zero air resistance).

Isaac Newton (1642–1727) and Gottfried Wilhelm von Leibniz (1646–1716) solidified a new method of knowing. Each in his own way and with their ideas combined convinced the Western world that it was possible to describe the universe objectively without reservations and remainders. Since, according to Newton and Leibniz, inert, immutable, universal "things" constituted the universe, everything had its place and science could define that place and describe its primary qualities and causes. The method was at first ill defined because it combined observations, deductions, inductions, creative insights, thought experiments, laboratory experiments, and mathematics—but the end result was the ultimate goal of universal laws. The proof of the method was that science could predict with mathematical precision where objects would be and even where they had been. An objective fit was possible; those who thought of themselves as scientists believed in it with their whole hearts and minds.

Immanuel Kant (1724–1804) played the decisive role in the establishment of the new methodology. He gave final expression to a new epistemology to meet the needs of a Newtonian universe. Kant was more aware than his predecessors of the constructive nature of the human mind. An intellectual fit required the union of two faculties: the faculty of understanding along with that which existed outside the human mind. Thus the "objects" of the world, when received through the senses, are made intelligible when conceptualized according to universal categories of pure understanding. Accordingly, truth became identified with the *procedure* of fitting together a priori mental concepts and things-in-themselves.[51] In Kant's own words, "understanding is itself the

51. Kant epitomizes the philosopher's basic mistrust of experience. He was certainly right that experience/observation cannot present to the mind things-in-themselves, because we can only know something as our mind constitutes it, not as a thing-in-itself. As a champion of this philosophical tradition, and in this respect a tradition that stands in tension with empiricism, Kant believed certainty would be possible only if there were con-

lawgiver of nature." Thus, nature cannot teach reason anything that reason does not itself order (Ormiston and Sassower, *Narrative Experiments,* 59). In place of Aristotelian ideal types or Platonic ideas or revealed truths, Kant established a "rational a priori." In a significant sense essentialism is transmuted from things-in-themselves (existing ontologically outside the mind) to a priori categories (existing epistemologically within the mind). Epistemology is likewise transmuted so that the procedure for knowing truly and certainly is no longer to discern (deduce, meditate upon) the mind of God but to construct from observation and experience. The triumph of method was no less than the crowning of epistemological certainty, as opposed to ontological certainty, predicated on the fit between the external world and its "objective" apprehension by the mind.

Evolutionary theory intensified the debate about methodology. Darwin's *Origin of Species* served to strike the last blow against what proponents of scientific method regarded as supernatural or ad hoc explanations. Darwin accepted the Newtonian world picture as well as the method of empiricism that integrated hypothetical formulation with observation and experimentation. There was, however, a fundamental difference, because Darwin was working with the world of living organisms. Here, Christian theology had found a place to hold out. The doctrine of special creation was still widely accepted, although the idea of a hierarchical chain was being challenged by various rival theories. As a scientist of his time, Darwin was caught in the tension of a mechanical model that would not translate into living systems. Darwin was nevertheless completely convinced that science was science because it utilized the same methodology throughout all scientific disciplines

There was considerable debate about Darwin's methodology. One of Darwin's staunchest defenders, T. H. Huxley, understood *Origin* to be "a mass of facts crushed and pounded into shape,

---

cepts or categories "in-themselves"; that is, present prior to experience. See the extended discussion in section 6.

rather than held together by the ordinary medium of an obvious logical bond; due attention will, without doubt, discover this bond, but it is often hard to find" (quoted from Hull, *Darwin and His Critics,* 32). It was not simply that Darwin's methodology was measured against Newton but that it was compared to the Newtonian myth and the unrealistic standards set by such philosophers as Mill and Herschel (see Hull, 27–31). Darwin found himself at a decided disadvantage compared to Newton, because evolution occurred so slowly and in such an indirect manner that no one had observed, nor was likely to observe, one species evolving into another. Darwin's evidence was statistical in nature and consequently inappropriate for shaping into universal laws. When paleontologist F. J. Pictet reviewed *Origin of the Species,* he wrote that he would not accept Darwin's deductions until he saw for himself the evolution of a new organ. Darwin could only sigh that Pictet, and critics like him, simply did not understand the character of the phenomena at hand. Darwin was, however, quick to draw comparisons wherever he could. If Newton was allowed to demonstrate the movements of physical bodies without being able to explain how gravity works, why should he have to explain how the first forms of life originated before demonstrating that species are related like branches of a bush? Darwin wished his critics would realize the unreasonable character of their remarks, since no one has yet seen a new species evolve or seen one divinely created.

Darwin was under no illusion that empiricism was strictly inductive, leading directly to assured results. He succeeded in opening the umbrella of empiricism to include living organisms because he had a clear understanding about the dialectical way that empiricism moves back and forth between hypothesis, deduction, and induction. For Darwin the more important requirement was to separate teleological questions (for what purpose?) from matter-of-fact questions (what does observation show us?). It is here that we find the greatest difference between Newton and Darwin. Newton felt obliged to function as a scientist with theological assumptions, although his assumptions had been narrowed and reduced; Darwin could not tolerate theists who would not keep their theology out of their science. Unlike his critics, such as his friend and codiscoverer of evolution Alfred Wallace, Darwin insisted on reporting an entirely consistent naturalistic story or none at all.

The scientific revolution that had its beginnings in the fifteenth century was now complete. Science was queen apparent because it had demonstrated a unique approach to knowledge that was superior to all other methods. The new method became the standard for knowing. The very term "scientific" was associated with, if not completely identified with, the "true" and "impartial." By the close of the nineteenth century, every discipline, including theology, tried to model itself on scientific objectivity. Scientific methodology had indeed become the conscious and unconscious communal definition of what is knowable and thus, according to Michel Foucault's definition, a new episteme. During this period of transition, science and theology were each confident that its methodology secured a kind of divine certainty; theology claimed the right to define what is real by what is true, while science asserted the right to define what is true by what is real.

The rivalry between science and theology was not, as is often construed, a choice between reason and faith, or the superiority of induction or deduction. Theology and science found it increasingly difficult to converse because their entire *epistemological* approaches to truth and reality were a series of contrasts.

Science	*Theology*
Mathematical equations	Theological propositions
Diagrams	Illustrations (manuscript)
Experimentation	Meditation
Demonstration	Inspiration
Curiosity	Piety
Fact gathering, observation	Interpreting the past, preserving traditions
Masters and possessors of nature	Wisdom and insight into the good
To discover and invent	To obey and glorify

This should not overlook the difference in the *ontologies* they each accepted.

Becoming	***Being***
The world as evolving	The world as created
What is factual	What is revealed

The new epistemology contained several important themes, including the democratization of knowledge, scientific "knowing" as constructed and assertive, scientific knowledge as cumulative and progressive, and the power of disengaged reason or objectivity. I will discuss each of them in detail, beginning with the democratization of knowledge.

Everyone's right to know dominates Bacon's *Novum Organum.* Bacon and Descartes argued that science would benefit the ordinary life; framed in such a manner, the argument was a thinly veiled criticism of theology's method of producing elite knowledge for a few resulting in no practical good. Common to all of the advocates of empiricism was their rejection of the medieval social understanding of a hierarchy of authority and a privileged locus of the sacred. The Protestant Reformation accepted and enlarged this sentiment, because its leaders understood that knowledge is power and power in the hands of the many is better than in the hands of a few. That each believer had the right to know was exemplified in such reformed teachings as the priesthood of all believers, salvation by faith alone, and freedom for individuals to interpret the Bible. There was a general social leveling inherent in the affirmation that the common life could be the locus of the sacred. Marriage and various secular vocations were imbued with new dignity.

The seat of authority shifted from the institution to the individual. The new foundationalism presumed that self-evident truths could be understood by anyone, regardless of their religious or political persuasion. The Declaration of Independence, for example, did not appeal to any other source of authority outside of itself. "The text, the narrative that is the declaration, conditions its own authority," because reason alone is capable of discerning the laws of nature (Ormiston and Sassower, *Narrative Experiments,* 56). Thus, there were self-evident truths obvious and clear to all, and there were other truths not quite so self-evident but nevertheless accessible if the right procedure was followed. Galileo found himself in conflict with the Church in large part because he believed the book of nature could be understood by reason without the help of an inspired text dependent upon an authority with a privileged source of knowledge.

One can hear in Galileo and Newton the continual appeal to common sense. Both petitioned the public through the art of persuasion, as if everyone could and should be convinced. "The Euclidean framework of the *Principia,* although occasionally forced, strikes home as a highly effective example of the use of *rhetoric,* the art of persuasion, in a scientific text."[52] There was little algebra in the *Principia.* Instead, there was ample use of the ancient language of geometry and proportion, albeit mixed with Newton's own invented mathematics. Even where the mathematics were foundational and few could follow the "proofs," most people were willing to accept what Newton said on faith, because there was a sense that anyone in principle could understand a Newton. Not so with theology, because theologians operated with some mysterious impenetrable method and their conclusions were never open to revision and corrections.

Another theme underlying the new epistemology was "knowing" as constructed and assertive. In his study of the seventeenth century, Amos Funkenstein examines the possible antecedents of "knowledge by doing." Behind the triumph of method was a sharply contrasting understanding of discovering new truth not by exposition of old texts but by doing or constructing. This new, energetic ideal of knowing stood squarely against

52. John Roche, "Newton's *Principia,*" in *Let Newton Be!* ed. John Fauvel, Raymond Flood et al. (Oxford: Oxford University Press, 1988), 49–50; emphasis added.

the old, contemplative ideal. Common to most ancient and medieval epistemologies was their receptive character; whether we gain knowledge by abstraction from sense impressions, by illumination, or by introspection, knowledge or truth is found, not constructed. Implicitly or explicitly, most "new sciences" of the seventeenth century assumed a constructive theory of knowledge.[53]

The self-assertive character of the scientific method, as Hans Blumenberg underlines, eradicated the value of watching the world from an attitude of repose or contemplation. The transition to a modern consciousness was marked by the shift to a "self-conscious curiosity." Nietzsche names "ruthless curiosity" as one of the epochal characteristics of the modern age. Religious authorities warned that an intemperate curiosity would invariably yield to a knowing beyond due measure. In Dante's Hell, for instance, one does not meet a heroic Odysseus but a reprobate who crosses the boundary of the known world with unbounded curiosity (*The Legitimacy of the Modern Age,* 338). The celebrated cases of Nicholas of Cusa (1410–64) and Giordano Bruno (1548–1600) are illustrative of the spirit's insatiability for knowledge and the price that had to be paid for challenging the medieval synthesis of theology and philosophy. Once people had presumed to peek behind the "realms of knowledge" or the "scenes of creation," a primal energy was released to know all things. Thus a resentment grew up against anything which impeded humankind's entitlement to know. The new sin became the indolence not to inquire aggressively of the world (*The Legitimacy of the Modern Age,* 235–37).

The new epistemology understandably viewed scientific knowledge as cumulative and progressive. At first, it was difficult to say how scientific knowledge was different from theological knowledge. There were distinguishing marks such as thought experiments, ideal conditions, practical demonstrations; but early

53. Amos Funkenstein, *Theology and the Scientific Imagination from the Middle Ages to the Seventeenth Century* (Princeton, N.J.: Princeton University Press, 1988), 298–99.

scientists like Kepler, Galileo, and Newton were content to consider themselves natural philosophers, a generalized term broad enough to include doing theology. Let us remember that as long as theology was sovereign, philosophy, science, and theology were almost the same occupation. Before long, however, common sense said scientific knowledge was different. It had specific and definable goals and it advanced steadily toward those objectives in a systematic way. People were awed by its ability to predict results and to do so with ever-increasing accuracy. In addition, scientific study became more encompassing. Systematically, year after year, a body of scientific knowledge accumulated. The library of scientific thought grew progressively until it can now be said that it doubles every ten years.

Contrary to popular notions, the new science was based upon probable reasoning, not on certainty. According to Ian Hacking, the logic of degrees of probability began to emerge with the Port-Royal School of thinkers in the second half of the seventeenth century, near the end of Descartes's life.[54] Theology was given to flaunting its superiority because its truths were steady, while science could only provide contingent or probable truths. When we are doing theology, Aquinas would say, we begin with first principles authorized by revelation and known through faith. Jeffrey Stout writes: "Divine authority is, for Aquinas and his contemporaries, the best conceivable warrant for belief" (*Flight from Authority,* 109). The very idea of probability as evidence would have been incredible to Aquinas; however, what probability offered, as did other forms of evidence, was a sense of partial truth steadily moving toward a fuller truth. In time this paradigm of steady advancement proved to be intoxicating.

While Hume's conclusion that all reasoning was provisional and incomplete was widely accepted, it was not easy to dislodge the conviction that theology was the keeper of eternal truths while

54. Ian Hacking, *Emergence of Probability* (Cambridge: Cambridge University Press, 1965).

science offered partial truths.[55] Science, however, had two deciding factors on its side. It showed progression, which meant that it was coming ever closer to a final truth. Probable truth also allowed for falsification, which in turn gave probable reasoning the air of being provable since it was also refutable. In the end, the traditional distinction between natural and supernatural seriously weakened theology's position. Most people made the choice for probable reason because authority could no longer maintain its claim to be self-authenticating. This was the first appeal to the proportioning of one's belief by weighing the evidence, and it was the first time that a growing body of evidence showed itself to be more and more probable until it became the more authoritative voice.

In the triumph of method, there were also the matters of tense and tenor. Theology's method was essentially retrospective, primarily interested in retrieval and interpretation. It offered little that was new. Theology was quickly becoming irrelevant. Science, on the other hand, was reaching toward a goal of building a new world. New theories preserved significant portions of the theoretical content of their predecessors. Thus, empiricism could assert that it functioned with a new kind of certainty, since Theory 2 was superior because it solved all of the problems of Theory 1, as well as some additional ones. For the "enlightened" person, science was exciting and creative. It was the court of final adjudication where more and more individuals wanted to serve.

The power of disengaged reason or objectivity was an additional theme in the new epistemology. Epistemology became the new ontology in that the process of knowing, in and of itself, became truth and reality. The radicalness of this turn toward procedure consisted of the belief that both reality and truth

---

55. Nancey Murphy concludes: "Thus Hume represents a great divide separating us from traditional theism, for in his work the consequences of the new probable reasoning were played out in theology. The burden of proof had shifted. Theology from Hume's day to the present seeks to defend itself not in the court of authority, but in the court of internal evidence." See Murphy, *Theology in the Age of Scientific Reasoning* (Ithaca, N.Y.: Cornell University Press, 1990), 12.

were not so much "out there," preexistent and ontological, but mentally constructed. Ironically, what saved this turn to the subject from becoming subjective was the belief in objectivity. Yet, it was belief in objectivity that became science's weakness in a postmodern era.[56]

Descartes himself is often mistakenly understood as giving us a theory of correspondence—namely, that truth emerges when the thinking mind makes a perfect fit between a priori ideas of the mind and things-in-themselves outside of the mind. The true power of rational thought was its capacity to discover and construct knowledge that had no prior existence. Both Bacon and Descartes were suspicious of the momentariness of immediate experience—reason was more disciplined, precise, and trustworthy. Both believed that the nature of things is evident not as embedded in the past, not even as directly and immediately experienced, but when things are withdrawn from their natural condition and exposed to an artificial mental reordering.[57] For the first time in Western civilization, truth was thought of as generated within the human mind—more significant than finding a proper fit or correspondence. A person could retreat to his room and cogitate a new truth about the world or about human nature. No wonder Descartes could utter, "I think therefore I am," and believe that a new era of knowing had dawned. In this regard, it is nearly impossible to underestimate the influence of the notion of scientific laws. Here is much more than Aristotelian observation or Platonic abstraction. This was proof that the human mind had the ability to bring together observation and rational thinking to formulate universal laws. The human mind had created something entirely new. Kepler, Galileo, Newton, and all who would be called scientific, strove to reduce understanding to foundational

---

56. Gerald Holton, *Einstein, History, and Other Passions* (Reading, Mass.: Addison-Wesley, 1996), 25–29.

57. Immensely helpful in this clarification of Descartes is Susan Bordo's book *The Flight to Objectivity* (Albany: State University of New York Press, 1987).

principles. Newton's *Principia* was so named because Newton believed principles would be his most enduring contribution.

The march toward the disengagement of the self from an ontological reference advanced on at least three fronts. First, the knowing self was not limited by or obliged to posit an ontological reference. When the world was stripped of its intrinsic features, it was freed to be engaged by the creativity of the human mind. Second, from the process of knowing, the disengaged self eliminated the Greek sense of self-knowledge or the Christian interplay of sin and grace. Third, such an epistemology severely reduced what could be considered real to the dimension of cause and effect.

Conclusion

The evidence of empiricism's triumph was twofold. The first, owing its most forceful argument to Descartes, was the transfer of certainty to the knowing process. Certainty arises because I have actively done something: of this I can be sure when I cannot be certain about philosophical speculations or revealed truths. The latter had no standards of evidence outside its own circular authority. Empiricism, as Bacon appreciated early on, posited standards of evidence that were not individual standards but standards that belonged to everyone who followed the proper scientific procedures of disengagement. The second proof was equally new. It was deceptively simple and almost tautological: what is true must be useful; what proves to be useful must be true. This would become the new pragmatism of the twentieth century. The process of coming-to-know was correct when it produced useful knowledge; that is, when it could predict and control a disenchanted world of matter. Theology was no rival for this form of truth, and philosophy did not offer truth with any immediate application. Curiosity was rewarded by building a better mousetrap. In the sixteen and seventeenth centuries, this was a window of opportunity too good to let go.

It was a monumental shift when Western intellectualism began to believe that the universe we live in could be known by the knowing self. This turn to the knowing self, as Charles Taylor elaborates, had its roots in Augustine but quickly distanced itself from him. Augustine took this path to show that God is found not just in the world but also in the individual's capacity for transcendence. We were encouraged to know ourselves because God was to be found in the intimacy of the self in the presence of God. From here the difference with the modern mindset widened. Augustine did not believe true knowledge of self or the world was possible without inspiration, revelation, and grace, because salvation required a method of coming-to-know unique to faith. His emphasis on the doctrine of original sin was a bar across the door of self-sufficiency. In the sixteenth and seventeenth centuries, the turn to the knowing self was recovered and made pivotal. The former safeguards, however, were discarded as impediments, the belief in sin and grace being foremost. For Protestants in general, and especially for Calvinists and Anabaptists, the community was an additional watchdog. A fundamental mistrust of the individual provoked John Calvin to believe there was no salvation outside the community, which was not so different from the Roman Catholic doctrine of no salvation outside the Church, if the Church is the priesthood of all believers. The very meaning of transcendence had also been changed. For Augustine, transcendence made it possible for us to know that there is a creator; for Descartes, transcendence made it possible for us to step outside ourselves with a disengaged perspective to generate clear and distinct ideas.[58]

More fundamentally, Taylor writes, the soul needed supplementation by grace in two ways. First, God calls humans to something more than the natural good. Beyond the natural virtues held by all decent, law-abiding citizens, stood the theological values of faith, hope, and love. Second, the human will is so depraved by sin that grace is required even "to make a decent attempt

58. For Charles Taylor's discussion of Augustine's turn to the self as radically different from the modern turn, see *Sources of Self,* 127–42.

at and perhaps even properly to discern the natural good, let alone to go beyond it" (*Sources of the Self*, 246). Locke looked upon human nature—as others had looked upon the cosmological world—as neutral, malleable stuff, waiting to be molded into a form that would bring universal happiness. Calvin, on the other hand, believed that human nature was certainly not a clean slate and that it therefore required the prevenience of God's grace in order for human nature to know and enjoy God. The natural good, however, had become a much thinner good. It had lost its ontological footing when the cosmological canopy was turned into mechanistic machinery. A lot was depending on the knowing self at a time when the self, while becoming more self-aware, was shorn of any meaningful place in the universe.

The triumph of method was originally a coalescence of objective concepts with objective things-in-themselves. However, this was predicated upon an epistemology that had purged the knowing self of personal, subjective values. The moral self was not exactly in danger of becoming extinct, but it was surely overshadowed by the "dignity of disengaged reason." In order even to continue speaking about the moral self, Kant found it necessary to separate duty and morals into a separate category of knowing that he called practical reasoning. This, too, was a disastrous turn. It deepened the impression that there were only two ways of knowing, one theological and speculative and the other empirical and pragmatic, deepening the separation of values from any ontological underpinning.

Operating with an ontological priority, theological notions of coming-to-know—such as inspiration and meditation—made sense. There had to be a preexistent order before God could disclose it; there had to be a Spirit in order to inspire, there had to be wisdom and goodness before the human mind could mediate upon it. The passive voice prevailed while theology was queen. Truth was a matter of conformity and harmony with something that was ontologically prior. Meditation, of course, was not meant to change what could not be changed, because the world had been created eternally good and had no need to be changed for the better. You meditated "day and night" in order to bend your life to the will of God. The benevolence and goodness of God was not something the individual created. Instead, it was found in a world that had been designed (Taylor, *Sources of the Self*, 124–26).

From this historical perspective one readily sees why the conversation between science and theology reached a low point at the end of the nineteenth century. Their rivalry had bruised egos and left one as a deposed sovereign over a shrinking domain and the other as a ruler over an expanding realm. As a discipline, theology was not sure whether to ride the train of science's methodological success or to find its own voice. But the latter meant a renewed engagement with the real when the ontological real was soon to undergo a relentless assault. Or should theology turn to experience or make several other turns to again be an authoritative voice? The following chapters will clarify these issues with further historical and philosophical inquiries.

Chapter 2

Where Have All the Sovereigns Gone?

There is another change that has had a good effect. It began about the turn of the century with new winds blowing in the natural sciences, and gradually the unbridled hubris that characterized science at the end of the nineteenth century gave way to a humility in the presence of reality, which is characteristic of natural science in our time when it is at its best.

—PROFESSOR ERIK OHLSSON in reference to Niels Bohr and Werner Heisenberg

It [modern science] kills God and takes his place on the vacant throne; so that henceforth it would be science which would hold the order of being in its hand as its sole legitimate guardian and be the sole legitimate arbiter of all relevant truth.

—VÁCLAV HAVEL, quoted from Holton, *Einstein, History, and Other Passions*, 35

[Theological] theories are shown to be scientific and therefore acceptable in the age when science lays down "where and in what sense we may speak of knowledge."

—NANCEY MURPHY, *Theology in the Age of Scientific Reasoning*, 211

- SECTION 4 -
SCIENCE THAT BECAME QUEEN BUT NOT A WORLDVIEW

The argument of this chapter is essentially twofold. Although science became the dominant mode of discourse in our modern culture, it did not effectively become a new worldview. Secondly, science capitulated from within. Science was certainly regarded as an autonomous queen, but only for a relatively short period at the end of the twentieth century. As the twentieth century ended, it became evident that science does not have the capacity to answer the perennial questions necessary to qualify as a worldview. This section delineates the ways that the integration of biology and evolution failed to establish a critical paradigm.[1] As queen-elect at the turn of this century, science also failed to hold together its marriage of ontology and epistemology. At first it seemed that evolution would supply both the missing piece (biology) and the critical dynamo (the paradigms of progress and becoming) to fashion a coherent account of the universe. The result was that science did not become an operative, logical way of being-in-the-world. Section 5 describes the nature of science's capitulation, and section 6 explains the dissolution of the bond between epistemology and ontology.

In the first chapter, I relied upon Langdon Gilkey's definition of what it means to be a queen. A queen possesses autonomy from other intellectual disciplines in theorems, data, and method, and is able to produce the kind of knowledge that is highly prized. Since science was unmistakably the new "queen," it should have supplanted a Christianized worldview with a nontheistic one. The so-called received understanding is that modern science tri-

---

1. In his comprehensive treatment of scientific revolutions, I. Bernard Cohen summarizes the impact of Darwin this way: "The Darwinian revolution was probably the most significant revolution that has ever occurred in the sciences, because its effects and influence were significant in many different areas of thought and belief. The consequence of this revolution was a systematic rethinking of the nature of the world, of man, and of human institutions." See his *Revolution in Science*, 299.

umphed as a new worldview, successfully supplanting religion as a coherent and objective joining of epistemology and ontology. There are several legitimate and cogent arguments in favor of this understanding. First, the Newtonian synthesis began as a perfect fit between hard things, such as atoms, and a methodology that generated universal principles. Second, the Industrial Revolution was for many people a new way of being in the world. Third, the received view points to the fact that science remains the dominant way of speaking about matters of truth. When you want to speak authoritatively, you try to sound scientific.

The argument, however, falters on many fronts. It rests upon the popular view that science is monolithic. It fails to distinguish properly between ontology and epistemology. It does not meet the standard of a worldview that requires science to provide satisfying answers to perennial questions concerning the kind of universe we live in and the kind of creatures we are. As an autonomous queen, science did not provide the resources for individuals and communities to distinguish what makes life worthwhile. Here I will principally argue that while science provided a superior method of knowing, the essential marriage or fit between epistemology and ontology was not sufficiently comprehensive and did not hold.

The seeds of modern science grew toward a worldview intended to displace the Christianized Aristotelian-Ptolemaic worldview. Bacon, Descartes, Kant, Hobbs, Hegel, and Marx envisioned a new world order. Bacon, for example, did not hesitate to claim universal application for the new scientific method of induction. When stripped of all pretense, Galileo was claiming nothing less than a universal form of knowledge. Kant attempted a systematic configuration of all knowledge according to the critique of pure reason. The three volumes of Marx's *Das Capital* were a comprehensive and critical system of self, nature, and society. Nevertheless, the question before us is whether a new worldview became operational.

The answer is a qualified but temporary Yes, and the most important evidence is the Industrial Revolution (c. 1760–1850).

Unquestionably, the changes associated with the Industrial Revolution, such as urbanization, mobility, and technological advances, required coping skills. The conceptual revolutions—the way we think about the structure and causation of the world—were immense. A new world picture situated the earth and its inhabitants in a vast and impersonal universe. Instead of looking up and being able to envision a sacred canopy stretching forth and mirroring the perfection of its creator, the human eye "saw" a clock-like order of planets and stars moving according to precise laws. On Earth itself, nature was deteleologized, its value transferred from a foundation for self-identity to a mine of raw energy. A new culture began to take hold as the consequence of understanding the world as a system of dead, inert particles moved by external forces. The mechanization of the world became a model for transforming political institutions and daily routines. The capitalist entrepreneur, free to create independent business ventures, replaced the sacred chain extending from king to peasant. In place of royal decrees and feudal stratifications, society was defined by free trade, economic efficiency, and independent nation-states. A new concept of the self as a rational master with unlimited potential displaced the self as a member of a close-knit, well-defined, familial structure. The natural rhythm of day and night was disturbed by the perpetual work of machines. Families were fragmented for the sake of making money. Waves of emigration were set in motion by hopes of a better tomorrow in an urban utopia of equal opportunity. God, the external governor and director, became more and more irrelevant in a world where humankind could harness the power of nature and penetrate the mysteries of the atom. In the modern era, belief in providence was a relic of a passive mind; progress was made by seizing the moment and taking matters into our own hands.

The Industrial Revolution required a new way of being in the world, but did science supply more than the technique for more efficient production? Historian of science Gerald Holton highlights what is inarguable: "[T]he methods of argument of science, its conceptions and its models, have permeated first the intellectual life of the time, then the tenets and usages of everyday life" (*Einstein, History, and Other Passions*, 43). The Newtonian synthesis generated a "storehouse of imaginative tools for thought," such as space, time, matter, law, causality, forces, verification, conservation laws, feedback, invariance, and complementarity. Holton, however, is unwilling to conclude that science operated as a worldview. To extrapolate Holton's argument, there is a difference between a worldview and a "storehouse of imaginative tools." Characteristically, Holton only mentions those tools that come from physics. There is of course a storehouse derived from biology, including chance, adaptation, natural selection, survival of the fittest, gradualism, mutation, cloning, and ecological balance. Did scientists or philosophers ever put these together in a coherent and integrated fashion?

While Darwin himself did not strive to create a new worldview, Herbert Spencer did. Nine years before Darwin's *The Origins of Species* appeared, Herbert Spencer had joined organic evolution with the social ideas of Adam Smith, Thomas Malthus, and David Ricardo. Spencer had been converted to Lamarck's theories of evolution, which held that traits could be accumulated and transmitted. Spencer was thus confident that the competition between individuals, population pressures, laissez-faire capitalism, and the inexorable march of evolution would eventually result in the upward advancement of civilization. Spencer would soon be joined by Darwin, Wallace, T. H. Huxley, and by liberal theologians like Lyman Abbott.[2]

This worldview, however, was short-lived. Spencer, Darwin, Wallace, and T. H. Huxley were at first enchanted with what Huxley called the picture that science draws of the world—"Harmonious order governing eternally continuous progress." Of the four scientists, Darwin remained the most committed to the explanatory power of natural selection as random and with-

2. See John C. Greene, *Science, Ideology, and World View*, 148–55.

out direction. He lived and died with nagging questions about the lack of meaning of a process that was at best opportunistic but blind. Wallace would part company with Darwin and Spencer concerning the qualitative difference between the human species and all other organisms. Humans were the superior species. Social and historical conditions forced Spencer to see the darker side of competitive struggle. In 1893 Huxley declared that he could see no way to ground social behavior in a non-law of chance occurrences and a law of survival of the fittest.[3] The one conclusion that somehow persisted is the call issued by the grandson of T. H., Julian Huxley, to fulfill our destiny "to be the agent of the world processes of evolution, the sole agent capable of leading it to new heights, and enabling it to realize new possibilities" (*Science, Ideology,* 165).

In *Darwin's Dangerous Idea,* Daniel C. Dennett finds creative ways to restate familiar but problematic themes. Two of Darwin's dangerous ideas proved to be a threat to physics as well as to theology. The first was the overthrow of essentialism, which depended upon timeless, unchanging, all-or-nothing things-in-themselves. A tight epistemological-ontological fit required the assumption that real things were being investigated, and in this scenario, the real had to meet a Newtonian definition. The living organisms Darwin discovered, on the other hand, were anything but timeless, unchanging, and definitive. There were stages of development, gradual adaptations, death, and speciation. Such powerful ideas as "descent with modification" and "natural selection" had no parallel in the inorganic universe of physics and astronomy. The methodology of these life sciences was distinctive. Classification and historical comparisons were

3. Greene, *Science, Ideology*, 158–89. Is this the beginning of the end of inferring an ethic of "should" from the "is" of the world? In Adrian Desmond's excellent biography, Thomas Henry Huxley has nothing less than a burning passion for a public education that teaches a science based upon principles of evolutionary theory free of a miraculous, cause-and-effect-nature. As Desmond notes, "We owe to him [T. H. Huxley] that enduring military metaphor, the 'war' of science against theology." See Desmond, *Huxley: From Devil's Disciple to Evolution's High Priest* (Cambridge, Mass.: Harvard University Press, 1994), xiii.

primary. For all who were operating with a Newtonian mindset, the methodology of biologists lacked the precision of mathematical formulation. The comparative-historical character of how to measure change was also disconcerting to scientists who tolerated only those standards established by physics.[4]

The second dangerous idea was an alternative way to answer questions such as why the world exhibits so much beauty, order, and design. The pre-Darwinian way was to invoke a designer following an argument that stated that where there is smoke, there is fire; where there is design, there is a designer (*Darwin's Dangerous Idea*, 24–28). Rather than invoking a teleological explanation, one could argue that order arises from a blind but patient process of gradual steps leading to greater and greater fitness and adaptation. In Dennett's words, Darwin turned Locke's mind-first principle upside down: "[I]ntentionality doesn't come from on high; it percolates up from below, from the initially mindless and pointless algorithmic processes that gradually acquire meaning and intelligence as they develop" (205). When placed side by side—the paradigm of a mechanical clock-like universe and a view of the world that was forever becoming—consolidation was impossible. Nor did the hard ontology of physics and the softer ontology of the burgeoning social sciences readily blend. While there was hope and even a presumption that the two kinds of sciences were sharing the same methodology, it proved to be a frustrating amalgamation.

One of the thorniest problems was the concept of time. Stephen Jay Gould reminds us that we in the Western hemisphere are dependent upon two metaphors of time. There is the direction of time's arrow and the nonlinear movement of time's cycle. The one metaphor makes intelligible the uniqueness of distinct and irreversible events, while the other constitutes the intelligibility of

4. See Ernst Mayr, *The Growth of Biological Thought* (Cambridge, Mass.: Belknap Press of Harvard University Press, 1982), 35–47, 201–8; David Hull, *Darwin and His Critics*, chap. 2.

timeless order and law-like structure.[5] The observation that we must have both was neither obvious nor reconcilable in the eighteen and nineteenth centuries. The sciences allied themselves with the uniformity and perfection of the cosmos. It was essentially a linear, a-historical view of time. Theology of the seventeenth and eighteenth centuries had created meaning by dropping miracles along a linear line—but it was anything but historical. Stephen Toulmin and June Goodfield also speak of a "science without history" (*The Discovery of Time,* chap. 2). Thus, both theology and the science of physics were confronted by a biological concept of time where extinction and new species could not be ignored.

Because physics had become the model for what science should be, the dominant paradigm was the concept of reversibility: a truly physical event had no direction because it was governed by causal principles that were the same then, now, and forever. In order for rival paradigms like evolution to be accepted, it was necessary for time to be extended by thousands, even millions, of years, as well as given depth and color. The longer extension of time was not a major difficulty for physics or biology, but the nature of change was. Loren Eiseley points out that there is one thing life does not appear to do: it never brings back the past. Unlike matter-in-motion, life seems to be traveling in a unique fashion in the time dimension (*The Firmament of Time,* 57). What now seems so easily discernible and commonplace, the observation of extinctions, was the same kind of anomaly as claiming that there were four moons orbiting Jupiter. The moons, however, were observable if skeptics would just look through Galileo's telescope and believe what they saw, while the extinction of a species defied direct observation. These two irreconcilable metaphors, time as universally the same and time as everywhere different and contingent, greatly frustrated the creation of a new worldview.

The consequence was that science could not inaugurate a coherent map of reality. Timothy Ferris points out that Galileo's

---

5. Stephen J. Gould, *Time's Arrow, Time's Cycle* (Cambridge, Mass.: Harvard University Press, 1987), 15–16.

telescope revealed a sky that has depth, but it was a cold, sterile depth (*Coming of Age in the Milky Way,* 89), not a place to find meaning and answers to perennial questions. On Earth, extinction signified a different set of laws that would have to account for waste, gradualism, struggle, new "creations," competition, and survival. Here time had character and movement but it was directionless. Newton could still believe in God's benign providence but Darwin could not, and their personal lives embodied two worldviews that refused to harmonize. Theology developed a split personality, with one profile fighting against evolution and holding onto the uniformity and continuity of a skyward universe, while the other reconciled with evolution and embraced change and progress as signs of the kingdom.

There are interesting examples of the twisting and turning that went on when physics and biology started to present conflicting views of the world. James Hutton, whose *Theory of the Earth* (1795) marks modern geology and the discovery of deep time, presented the earth as a self-renewing world machine endlessly repeating cycles of decay, erosion, settling and building up, uplifting and then returning to decay and erosion. The earth was alive but has in his own words, "no vestige of a beginning,—no prospect of an end" (*Time's Arrow,* 63). On one level, this is an extension of Newton's cosmos with the expulsion of all first causes. This was an a priori presumption on the part of Hutton. While change or deep time was allowed to penetrate the earth, it was mechanical and not biological. Seven short years later William Paley's *Natural Theology* (1802) represents an odd mixture of two worldviews trying to merge. His book begins with a mechanical illustration: a watch that begs you to answer how it was made, if not by a designer. This watch, however, has unusual properties: "the unexpected property of producing in the course of its movement another watch like itself" (*Natural Theology,* chap. 2). Paley did not seem especially bothered by an argument that explains one watch as "in some sense" the maker of the other watch. As his argument develops, almost all of Paley's examples of design are drawn from the world of nature (eye, muscles and joints, human frame, circulation of the blood, and so on). Only at the end of his argument, and before a theological treatise on the nature of God, does Paley briefly take up astronomy and reveal his rationale: "My opinion of Astronomy has always been, that it is *not* the best medium, through which to prove the agency of an intelligent Creator" (*Natural Theology,* chap. 22). If you were not aware of its publication date, you would think *Natural Theology* was a refutation of a Darwinian, atheistic, no-need-of God theory of evolution. Instead, it is a splendid illustration of a theologian caught in the transition between two worldviews.

The world described by Charles Dickens was qualitatively different than John Milton's, yet they both have a narrative structure. William R. Everdell describes the first moderns as those who profoundly changed the landscape of conceptual thought that reality is—discrete, atomistic, and discontinuous, contrary to the narrative seamlessness of nineteenth-century thought (*The First Moderns: Profiles in the Origins of Twentieth-Century Thought,* 9–12). In Dickens we are taken from Milton's pristine cosmos and plunged into the dirty, sooty microcosm of the city; here we engage a modern, secular way of being in the world. Picasso, Joyce, Dedekind, Boltzmann, Einstein, Seurat, and Picasso are modern thinkers precisely because they regarded Dickens's world of plot with beginnings and endings as not the real reality. Poems without meter, sentences that flow like a stream of consciousness, pictures composed of dabs of pigment, unpredictable particles that only give the illusion of statistical continuity—these are the defining art forms of modern science.

With the loss of continuity and a general thinning out of the landscape, another phenomenon was taking place. In his provocative book *Disappearing through the Skylight,* O. B. Hardison Jr. charts the short route from Charles Darwin's *Descent of Man* to D'Arcy Wentworth Thompson's *On Growth and Form* (1917) and concludes, "the sense of tragedy—of human drama with human consequences—is replaced by generalized appreciation of the beauty of form" (*Disappearing,* 103). Darwin still possessed a love for the particularity of each object, defined by its specific niche in time and place and yet part of the grander scale of life ascending from the simple to the complex. Thompson's vision, on the other hand, has only the "delight in the discovery of mathematical regularities" and, in that, loses the sense of plot and drama (*Disappearing,* 41–44). As the hard sciences came to dominate, universal laws of uniformity and sameness covered over the " 'storied fight' from matter to life, life to mind, mind to culture, and culture to spirit, a fight through good and evil" (Rolston, *Science and Religion,* 275).

Modernity required that meaning be found in a different place. One place to look—and if it was not the only place left, it was so pressing that it could not be ignored—was the study of humankind. In reviewing the major historical shifts toward this study, Frank Baumer writes this about the eighteenth century: "To put it another way, anthropology, that is, the study of man, or mankind, became the new queen of the sciences, displacing natural philosophy, which had been so all-absorbing in the seventeenth century, as well as theology, the old queen of Christian culture" (*Modern European Thought,* 160).

Baumer's analysis begins with Alexander Pope (*An Essay on Man*, 1733–34) and David Hume (*A Treatise of Human Nature,* 1739). But everywhere one looks there is the sense that Pope's famous couplet—"Know then thyself, presume not God to scan; The Proper study of Mankind is Man"—was a prophetic epigram that starting with the mind of God was no longer necessary. Enlightened metaphors and images were about to dominate: moral man (Lord Shaftesbury), rational man (Rousseau and Kant), economic man (Adam Smith), and perfectible man (Locke). As Baumer follows the history of this turn to the study of man, it does not abate with the next century: it only changes from essentially static concepts to historical ones. While Newtonian physics dominated, the knowing mind was modeled after the camera, a machine, and eventually, the computer. The study of living organisms, however, could not follow along. Some time later Werner Heisenberg was one of the first physicists to recognize that "even in science the object of research is no longer nature itself, but man's investigation of nature." Heisenberg, however, had in mind a science where the individual is inseparable from the discerning subject. A theory of evolution, more broadly formulated as belief that becoming is the essence of existence, required more than the individual to be included in one's epistemology; it demanded an understanding of human beings as part of the universe. Theology, which by now had become a lame-duck queen, was essentially ignored, but it left behind a tradition much richer and deeper than any that could be created by another discipline.[6] Theology was challenged to rethink and restate its response to how human beings know, in order to understand who they are "in the great scheme" of things.

The nineteenth century, therefore, closed with a number of enduring predicaments. Newtonian physics was the idealized model for doing science, while evolution was the more comprehensive concept for interpreting history and life. Physics and biology as representatives of the hard and soft sciences were different ways of doing science; how different was recognized only as science itself matured and gained a historical perspective. In spite of the enormous leaps in knowledge, only minimal bridges connected the world of living creatures and the atom. We still struggle to find standards of truth that fit well the course of physical events, historical events, natural events, human development, and the moral life. This disjunction between physics and biology was one reason, but not the only one, for a disjointed account of a discontinuous world.

---

6. I have in mind Karl Barth's blistering response to Enlightenment philosophers who thought they knew what constituted a proper study of man. See his *Protestant Thought from Rousseau to Ritschl* (New York: Harper & Row, 1959), passim.

– SECTION 5 –
THE NATURE OF THE CAPITULATION

In *Einstein, History, and Other Passions: The Rebellion against Science at the End of the Twentieth Century* (1996), the respected historian of science Gerald Holton argues that the downfall or derogation of science is due to a revolt, a romantic revolt. Thus, a shift has taken place whereby science no longer sits so high and mighty on the throne of human knowledge.

"A chief object of this countercultural swing," Holton writes, "is to deny the claim of science that it can lead to a knowledge that is progressively improvable, in principle universally accessible, based on rational thought, and potentially valuable for society at large" (*Einstein,* 3). This Romantic Rebellion, as Holton identifies it, is essentially the long-standing tradition of disenchantment with science represented by the classic Romanticism of Goethe, Schiller, Byron, and Blake; the Russian literary giants such as Turgenev, Dostoevski, and Tolstoy; along with images of dysfunctional scientists from Dr. Frankenstein to Dr. Strangelove; and its political manifestation in Mao's Cultural Revolution in China. This rebellion has consistently waged war against the optimism of inevitable progress based upon rationality and objectivity by endowing the romantic self with individualism and freedom. Holton believes that this Romantic Rebellion threatens to emasculate science and to rob Western civilization of its most potent "truth-seeking and enlightening process in modern culture" (*Einstein,* 4).

On one level the argument is whether Holton overestimates the potency of this counter-movement to dethrone science. His argument is considerably strengthened if one considers postmodern deconstructionism and fundamentalism, especially as the latter has developed into various forms of nationalism, as further manifestations of this counter-culture. I seriously doubt this Romantic Rebellion is capable of dealing science a fatal blow, and I hope, as Holton does, that we have not reached a critical historical moment when we must choose between two cultures: one rational and the other passionate, one empirical and the other spiritual (*Einstein,* 29). Regardless of any overestimation, Holton makes an important contribution toward explaining why science does not rule as a sovereign. I have chosen to speak instead of science's capitulation and in so doing to analyze the situation from quite a different perspective. Capitulation, instead of rebellion, implies an internal dynamic and I will consider this dynamic as moving from a hard to a soft ontology and the historization of science itself (below). One also cannot ignore the historical events of the twentieth century that by any account is a record of two World Wars, the Holocaust, Hiroshima, genocides, and other atrocities.[7] There were also those particular historical events that rattled our confidence in the technology that is science: the *Titanic,* Three Mile Island, *Apollo 13,* the *Challenger,* the Hubble telescope. By scale and intensity we had to consider the shortcoming of scientific "progress," the unpredictable nature of nature, and the irrational and evil behavior of human beings. I will also be discussing the proclivity within science toward overpromising and overreaching. To what extent this proclivity is a fatal flaw within science, perhaps even an unconscious ideology, leads us back to Holton and a series of essays by Isaiah Berlin.[8]

Holton, along with other historians, is interested in the connection between the ascendancy of science and the horrors of the twentieth century, especially as they were a manifestation of totalitarianism. In this regard he refers to Berlin, who asserts that science and tyranny are somehow intertwined—"that the development of the modern natural sciences and technology may, through the reactions against them, have unintentionally and in-

7. Jonathan Glover has written such an account in *Humanity: A Moral History of the Twentieth Century* (New Haven, Conn.: Yale University Press, 2000).

8. Isaiah Berlin, *The Crooked Timber of Humanity* (New York: Vintage Books, 1959), 207–37. Holton is also following Berlin in his analysis of a romantic rebellion intended to wage war upon the notion of objectivity and the worldview it spawned.

directly contributed to the rise of 'totalitarian tyrannies.' "[9] While Holton is not prepared to endorse any easy or direct link, neither does he entirely reject the suggestion by Berlin that there might be an intertwining. He finds in the Czech poet, playwright, and president, Václav Havel, a contemporary embodiment of this rebellion. Holton sees in Havel another popular indictment leveled at science, namely, that technology does more harm than good. In a frequently quoted essay, "The End of the Modern Era," Havel writes:

> The modern era has been dominated by the culminating belief... that the world... is a wholly knowable system governed by a finite number of universal laws that man can grasp and rationally direct for his own benefit.... This in turn, gave rise to the proud belief that man, as the pinnacle of everything that exists, was capable of objectively describing, explaining and controlling everything that exists, and of possessing the one and only truth about the world. Traditional science, with its usual coolness, can describe the different ways we might destroy ourselves, but it cannot offer truly effective and practicable instructions on how to avert them. (Quoted in Holton, *Einstein,* 33)

In this essay and other writings, Havel leaves no doubt that he understands there to be a direct connection between modern science and Communism as the perverse end of believing the world is objectively knowable, for if the latter is true then the forces of nature can be objectively managed. Holton will not draw the connection this tightly, but he does believe there was *"the ominous joining in the twentieth century of the extremes of a Romantic Rebellion with irrational political doctrines"* (*Einstein,* 29–30, italics in original). This, as he points out, was evident in the

9. The quotation is Holton's summary of Berlin's "eloquent and subtle analysis." See *Einstein, History, and Other Passions,* 26; italics in original.

Cultural Revolution in Mao's China.[10] But further analysis is required.

If science did not become a ruling sovereign, as I have argued, and if neither philosophy nor theology or science provided a coherent worldview at the beginning of the twentieth century, then a vacuum existed. Into this vacuum rushed dictators, ideologues, utopian dreamers who set about creating their own worldviews that we know as socialism, Communism, Marxism, nationalism, capitalism, and so on. What these "isms" had in common was the faith that one can rationally create a new society. Marx and Engels epitomized this belief but it is Mao Zedong's "Great Leap Forward" that exemplifies an attempt at utopian engineering. Jasper Becker's *Hungry Ghosts* (1996) documents how Mao's blueprint for a future when there would be a surplus of food—truly a utopia for a people where food was continually scarce—went wrong. Mao rejected Western scientific methods in favor of a science championed by the Russian scientists Lysenko, Michurin, and Williams under Stalin. Tragically, Mao intermingled ideology and scientific agriculture principles as if that would make no difference. Mao assumed that science should be common sense, capable of producing instant results and carried out by anyone. Close planting and deep plowing were two favorite techniques. The result was the death of more people than in Stalin's purges and the Holocaust together. Certainly "irrational political doctrines" fueled the various "isms" of our century but what incited them even more was the unbridled confidence in science as the means to remake society.

So it could be argued that it was not science but a pseudoscience that precipitated failure. That would be only partially correct. Political leaders co-opted science for their own gains. They mistakenly believed scientific principles from the hard sci-

10. Holton's other historical examples are not as convincing: the anti-scientific sentiment in the USSR, and sentiment against Aryan science in Nazi Germany. Even though this sentiment existed it made little headway against the belief that science provided *the* way to shape the future.

ences could be easily transported to the social sciences in order to build a new society. Undoubtedly, it took autocrats, empire-builders, and idealists to believe it was their turn to rule as a king. But it was also inherent in science to promise the world and those promises became grand schemes, final solutions, the superior race, the war to end all wars.[11]

Despite the enduring presence of a romantic revolt, science is still generally regarded as our best hope to construct an idealized future, but since the early part of the twentieth century a fundamental change has occurred. Science is closely watched by skeptics and cynics with a wary eye toward idealizing science. Scientists themselves have become more watchful. In *The Shadow of the Bomb,* S. S. Schweber has written how physicists became deeply reflective after creating the first atomic bomb. It quickly became apparent, he observes, that the way science was to be done was unalterably changed. After 1945, "it became ever more difficult to draw sharp boundaries between pure science and applied, between applied science and technology, and to insulate 'pure,' basic science from both its consequences and its requirement of societal support." This was a decided shift from the mentality at Los Alamos, for here there was an operative belief that pure knowledge is always good (*In the Shadow,* 19–20). Thus, science did not get its comeuppance so much because of a romantic or postmodern revolt. Rather, science no longer rules as queen because there is a degree of self-acknowledgment within science and a general skepticism from without.

Three general observations are in order as we move this discussion forward. First, remember that theology was dethroned by a new and powerful rival. Science, on the other hand, capitulated from within rather than by a forced surrender. By the end of the nineteenth century, science was queen "for a day," because

---

11. Glover draws our attention to the mentality of war in the twentieth century as it became ever more destructive and encompassing (e.g., the death of so many civilians). See *Humanity,* part two ("The Moral Psychology of Waging War") and part four ("War as a Trap").

it owned and utilized the one methodology that had proven to be cumulative. Throughout the twentieth century, there were other contenders and rivals; however, they were not out to displace science but to co-opt it.

Second, the time span we are about to examine is relatively short, surprisingly so: the core of the capitulation of science took place from 1900 to 1950. As with all historical periods, this one can be extended backward to its roots and forward to its consequences—about thirty years each way. Compare this short time with the hundreds of years that it took to dethrone theology. Because of theology's all-inclusive character and its great adaptability, it ruled both as queen of all human knowledge and as a worldview or framework for all of life; by contrast, science was queen only of a narrower spectrum of technical or useful knowledge. Nevertheless, one can only be astounded that the Newtonian synthesis was centuries in the making, but the radical displacement and fragmentation of that synthesis was only decades from beginning to end. Whether we account for this by the sheer speed of scientific advancements or by a Kuhnian analysis involving paradigm shifts and the importance of anomalies, the comparison between science and theology is striking and gives reason to suspect that the dynamics of capitulation and of dethroning were very different.[12]

My third observation concerns the not-so-subtle difference between power and authority. Theology's reign ended when its authority was abrogated by science and its power waned. By contrast, the power of science to influence is almost omnipresent but its authority has become compromised. Just ask yourself how you would feel about electing a proclaimed scientist to the White House; would that give too much authority to what is already

12. The reference to Thomas Kuhn is to his immensely important *The Structure of Scientific Revolutions,* 2d ed. (Chicago: University of Chicago Press, 1970). Kuhn's book and his discussion of paradigm shifts has stirred considerable controversy. See *Criticism and the Growth of Knowledge,* ed. I. Lakatos and A. Musgrave (Cambridge: Cambridge University Press, 1970) and H. Floris Cohen, *The Scientific Revolution: A Historiographical Inquiry* (Chicago: University of Chicago Press, 1994).

too powerful? The reason that we do not trust science to rule absolutely is rooted in our recent experience: scientists continually assure us that they can manage nature when they cannot. Chernobyl and the Challenger are jarring symbols that we are not masters of all that we survey. Auschwitz is that monstrous tragedy that proclaims: "We who have the power to manipulate and to shape nature are ourselves out of control." Since this is our legacy, and it can be written backward as far as we can remember, it is better to build a safety wall between power and authority.

To understand the capitulation of science in more detail, three themes are worth considering: the emergence of a softer ontology and the loss of objectivity, the historicization of science, and the nature of science to overreach and to overpromise.

The Softer Ontology and Loss of Objectivity

The softer ontology of modern physics is my way of describing the shift that took place beginning at the end of the nineteenth century. The conversion from a hard ontology to a soft ontology is a cornerstone in the transformation from a modern to a postmodern mentality. The perfect marriage between methodology (experiment) and ontology (a world that can be cut open and manipulated) of modern science was predicated upon reality being picturable and fully describable. The Newtonian ontology of discrete objects and events occurring within a framework of fixed space and time is just what an epistemology bent on finding objectivity would need to find. A hard ontology meant hard truth claims—an objective universe measured with an objective eye. Thus, science could claim that it saw the world as it really is.

One fault line that ripped apart the hardened crust of science's reign of method began in the golden age of quantum physics (1905–27). There was scarcely time for science to enjoy its triumph before it had to begin again with a new foundation. In a short time span, the three great revolutions in physics took place: the Maxwellian revolution (the *Treatise on Electricity and Magnetism* in 1873) and the great revolutions in quantum mechanics and relativity. As a result, neither the Newtonian synthesis nor Social Darwinism could endure without being significantly modified. Theories of simple causality gave way to explanations based on relativity, probability, evolutionary biology, and historical genetic programs. The splitting of the atom (the atomic era), the decoding of DNA (the genetic engineering era), and advancements in computers and information transfer (the era of cybernetics) left no doubt that a postmodern science would be a different ball of wax.

The subsequent revolutions inherent in relativity and quantum mechanics meant that something had to be revised—and that something was the nature of the fit between epistemology and ontology. As science was compelled to use relational language to describe matter, it was also compelled to reevaluate its methodology of objectivity. The fuzziness of objects, the indeterminacy of cause and effect, the hierarchical nesting of theories, the blurring of distinct lines, resulted in a reluctant but definite affirmation that the endemic richness of reality means we cannot know it as it is. The distortion of gravity on space-time requires a non-Euclidian geometry; therein, the universe loses its picturability. Because there are no simple one-to-one cause and effect happenings but only interrelated, interdependent events, the universe is incessantly successive and emergent. Objectivity is invariably compromised since every measurement influences what we measure. In short, the universe is ambiguous and evolutionary and therefore our understanding of it must be interpretive, contextual, and historical.

One of the hallmarks of postmodern and soft ontology is the persistent faith in methodology, even though the ontology of the universe is always in a state of flux. Romantics or romantic-like jabs will therefore have limited success, because they do little to undermine the faith that scientific methodology will eventually get it right. Of course, this leaves unanswered what it is that science will get right. To snipe away at the deleterious effects of technical progress is not very effective, since we are willing to swallow the pill of necessary mistakes along the way.

The Historicization of Science

Throughout most of their history, Western science and philosophy were committed to demonstrating the congruence of pure reason and ideal structures; such was the foundation of Aristotelian essentialism and Euclidian geometry, as well as of modern forms of positivism and realism. Natural philosophy, natural theology, and the empirical sciences were united in the common task of describing the essential nature of things, discovering ideal forms, giving unambiguous and complete explanations of ultimate reality. Although science attempted to ward off any suspicion of subjectivity, ultimately it could not fend off the twin nemeses of language and history. Stephen Toulmin, reflecting on why Thomas Kuhn's book *The Structure of Scientific Revolutions* has been so influential, makes the important comment that "Kuhn completed the historicization of human thought that had begun in the eighteenth century, and so finally undercut older views about the 'immutable' order of nature and human knowledge."[13] Toulmin concludes:

> If natural science no longer makes any claim to permanent or fixed ideas—if the sciences are always liable to run into "crises" that force us to dismantle them and reconstruct them on new foundations—then physical theory, which had for so long remained the last stronghold of intellectual absolutes, is finally engulfed in the same historical flux as the human and social sciences. (*Paradigm Change*, 236)

There are degrees of willingness on the part of individual scientists to surrender a universe composed of identifiable things with

13. Stephen Toulmin, "The Historicization of Natural Science: Its Implication for Theology," in *Paradigm Change in Theology*, 233, 236.

intrinsic attributes that exist independent of the observer.[14] What cannot be denied is a fundamental shift in ontology requiring an epistemology that takes into account the historical and social context in which science is done, the historicity of all human thought, and the role of language. The fact that most universities now have a discipline devoted to the history and philosophy of science is sufficient evidence that science sees itself as a hermeneutical and a historical enterprise.

It is through language that thought is embedded in history. The acceptance of this terse statement was a long time coming, because Western philosophy was determined to demonstrate just the opposite. For all their differences, Descartes's clear and distinct ideas and Kant's a priori consciousness both intended to provide the basis for a trans-historical foundationalism, in which truth and the world could be known by the natural light of reason. Newton's *Principia* inspired philosophers to apply the principles of objectivity to language. The quest for a scientifically precise language reached its pinnacle when the Vienna Circle initiated "the linguistic turn" in the first part of the twentieth century.[15] Bertrand Russell and the younger Ludwig Wittgenstein were the epitome of this turn in their proposal for a formal marriage of pure mathematics and symbolic logic.[16]

Theologian David Tracy summarizes the shift that has taken place, regarding the inadequacy of a positivistic understanding of language.

---

14. See Charles Coulston Gillispie, *The Edge of Objectivity* (Princeton, N.J.: Princeton University Press, 1960) and Wilfrid Sellars, *Science, Perception, and Reality* (New York: Humanities Press, 1963) for a discussion of the vanishing notion of objectivity. John L. Casti has a good general discussion of the "reality problem" or why the existence of independent entities is no longer a taken-for-granted postulate. See his *Paradigms Lost* (New York: Avon, 1989), especially chap. 7 and the conclusion.

15. See Richard Rorty, ed., *The Linguistic Turn* (Chicago: University of Chicago Press, 1967) and *Objectivity, Relativism, and Truth: Philosophical Papers*, vol. 1 (Cambridge: Cambridge University Press, 1991).

16. See William Barrett, *The Illusion of Technique* (Garden City, N.Y.: Anchor, 1967), chap. 3.

> With science we interpret the world. We do not simply find it out there. Reality is what we name our best interpretation. Truth is the reality we know through our best interpretations. Reality is constituted, not created or simply found, through the interpretations that have earned the right to be called relatively adequate or true. In science, language inevitably influences our understanding of both data and fact, truth and reality. (*Plurality and Ambiguity*, 48)

Tracy clarifies the new understanding of language by contrasting it with both the romantic and positivist stances, which assume that language is purely instrumental; that is, language comes into play after the fact of cognition and discovery. The romantic uses language to express some deep, nonlinguistic truth inside the self. The positivist uses language to express and communicate scientific facts. "Language," writes Tracy, in both the positivist and romantic readings, "is secondary, even peripheral, to the real thing. The real thing is purely prelinguistic: either my deep feelings or insights from within or my clear grasp of clear, distinct, scientific facts" (*Plurality and Ambiguity,* 49). Language, then, is simply a method of bringing to expression what is already present.

The alternative consensus is that we do not first experience or understand some reality and then find words to name that understanding. Instead, we understand in and through the language available to us, including, for example, the particular languages of theology and science. We should concede that even though there are nonreflective experiences, there are no uninterpreted or unschematized ones. Graham Ward concisely summarizes the two contrasting models for language:

> The communication model (where words adequately represent and communicate their objects) cannot be accommodated within a model of language which understands words as constructing the reality of objects. Similarly, the semiotic model of language (which emphasizes the ineradicable me-

diation of a "meaning" which forever lies beyond it) cannot accommodate the possibility of unmediated, direct disclosure. (*Barth, Derrida and the Language of Theology*, 30)

The communication model of language understands this movement as linear; that is, it accepts the possibility of making a direct correlation between words and reality. The construction model postulates a dynamic movement as language shapes and intensifies experience. For the latter, the spaces between viewing, conceiving, and expressing require that there can be no immediate or direct transfer of meaning.

Since the process of coming-to-know happens in and through language, there is an inescapable social and historical character of all understanding that takes place through language.[17] Both theology (by way of revelation and inspiration) and science (by way of empiricism) believed that as sovereigns they were above the flux of history and the conditioning of language. Such, however, is not the case: they share a compelling commonality, namely that meaning, whether it is the meaning of facts or the meaning of historical events, is created and understood through language.

Science Overpromises and Overreaches

My third observation about the capitulation of science can be described as overpromising. G. K. Chesterton said in effect that the only empirically provable doctrine of Christianity is original sin. Putting aside the thorny question about the interpretation of "original," Chesterton had in mind the ever-larger scale of war

17. In general a considerable level of historical consciousness has been raised by scholars who have demonstrated the historical context of all scientific paradigms (Kuhn, see n. 12), the historical character of all scientific arguments (Stephen Toulmin, *Foresight and Understanding: An Enquiry into the Aims of Science* [Bloomington: Indiana University Press, 1961]), the creativity of scientific discovery (Karl Popper, *The Logic of Scientific Discovery* [New York: Harper & Row, 1968]), the participatory nature of the universe (John Wheeler, *At Home in the Universe* [New York: American Institute of Physics Press, 1994]), and how the problem of observations has been eclipsed by the problem of meaning (Susanne K. Langer, *Philosophy in a New Key*, 3d ed. [Cambridge, Mass.: Harvard University Press, 1956]).

and destruction. While science is not to be blamed for creating a more evil human being, science did lay the groundwork and create the mindset for a new hubris rooted in the belief that human beings will act rationally, can be educated progressively to improve, and have the right to seek truth at any price. In other words, science has ignored what every religion teaches about the ontological nature of being human. Two studies by William Leiss and Jacques Ellul explore the dimension of sin as overpromising and overreaching.

Leiss begins with the argument that science was a favored instrument in a grand design that came to be known as the domination of nature. Incorporating a study of Jerome Tavetz (*Scientific Knowledge and Its Social Problems* [1971]), Leiss claims that an "ideology of science," which is no less than a collective social enterprise held by scientists and nonscientists alike, functioned to promise the human species prosperity and fulfillment by "gaining complete control over the forces of the natural world, appropriating to the full its resources for the satisfaction of human needs" (*Domination of Nature,* xi). The mastery of nature, however, was not the dominant dynamic.

> In the false hopes fostered by the idea of "conquering" nature was concealed one of the most fateful historical dynamics in modern times: the inextricable bond between the domination of nature and the domination of man. How the latter emerged as an unintended consequence of the former is one of the most difficult riddles which we have been called upon to resolve. (*The Domination of Nature,* xiii–xiv)

What actually happened, Leiss argues, was the domination of human beings under the guise of managing nature. Leiss recognizes both an internal and an external dynamic. The former springs from Cartesian philosophy, "where the ego appears as dominating internal nature (the passions) in order to prevent the emotions from interfering with the judgments that form the basis of scientific knowledge" (152). The external dynamic was fired by Bacon's conviction that "men could fundamentally alter the material conditions of existence," thereby breaking the tyrannical hold of despair over their lives (175). Brought together as an internal and an external force, the ideological dynamic to master and dominate is well known in the last century.[18] When Leiss asks "what is the historical dynamic that spurs on" this drive to master and dominate, his answer is social conflict, which he says can be remedied by reducing greed and lowering desire for more and more goods (152–65). His concluding statement identifies where his hope lies: "Liberation is equivalent to the nonrepressive mastery of nature, that is, a mastery that is guided by human needs that have been formulated by associated individuals in an atmosphere of rationality, freedom, and autonomy" (212).

Few would deny that the domination of nature, the domination of human beings, and social conflict compose a single fabric, but does Leiss's understanding of human nature have sufficient ontological depth? At stake is the perennial question concerning human nature. Once "collective rational control is established," he writes, "technology will be liberated from its all-too-effective service in the cause of human conflict" (161). The Romantics, as Holton recognizes, challenged the very assumption that rational self-control is possible and understood better than most the danger lurking behind the promise of utopia in exchange for a collective mind or purpose. Christian theologians must protest as well, because they understand self-control and the acceptance of limits as one of the gifts of living by the Holy Spirit.[19]

---

18. Morris Berman makes a similar observation concerning the importance of Galileo's understanding of the relationship between theory and experiment; namely, "there can be a fundamental link between cognition and manipulation, between scientific explanation and mastery of the environment." It matters less whom we credit as long as we recognize that science was about to embark upon a new and bold venture, because the mind has the power to predict and change the nature of nature. See Berman, *The Reenchantment of the World,* 54.

19. I am thinking particularly of the Pauline theology with its emphasis on self-control as a manifestation of the lordship of Christ and "the fruits of the Spirit"; see Gal. 5:23; 1 Cor. 7:5–9, 9:25; 2 Tim. 1:7. Conversely, a sign that one is still in bondage to sin is a life that is out of control or where passions "run wild"; see Rom. 1:26–27.

Is science exonerated too quickly with statements to the effect that the domination of nature ought not to be identified with scientific and technological progress (*The Domination of Nature,* 175)? Science was born and nurtured by the paradigm of manipulation, objectivity, and rationality. Its association with these is one reason why it is has fallen from grace, but it has not fallen so far that science has lost its promise value. Surely, manipulation, rationality, and objectivity have a life apart from science; in reality, they are so entangled in our scientific culture that they are not easily separated. It is no accident that scientific methodology is invariably applied with the promise of a better tomorrow.

Taking a very different tack, Jacques Ellul is more suspicious of what the ideology of science has wrought. In his *The Technological Bluff* (1990), Ellul defines technique as "the totality of methods of rationality arrived at and having absolute efficiency in every field of activity" (xi). Technology then becomes the specialized application of science to a particular problem. The essence of the bluff, then, is the belief that science can do all things using the all-powerful tool of technique. This belief, however, has been severely undermined by several counter beliefs. First, all technological progress has its price; second, at each stage it raises more and greater problems than it solves; third, its harmful effects are inseparable from its beneficial effects; and fourth, it has a great number of unforeseen effects (*Technological Bluff,* 39).

Ellul supplies the reader with numerous illustrations. For example, the discovery and understanding of how to release the energy of the atom is one of the premier examples of the ambivalence inherent within technological progress. We have mixed feelings whether or not the splitting of the atom has made the world a better place. There are those who say we cannot live without atomic energy and those who say we cannot live with it. Some believe that science will find a satisfactory way to handle nuclear waste, but it is one of those significant and intrinsic unknowns that cause others to have second thoughts. Weighing down one end of the scale is an irrational stockpile of hydrogen bombs and

Chernobyl; the counterweight includes all the beneficial applications of nuclear medicine. The splitting of the atom generates ambivalent feelings because science itself cannot help but reveal a potent universe beyond human mastery.

Waste is another inescapable but unforeseen consequence of technological advancement. At first we did not consider waste to be much of a problem. We did not consciously decide to weigh the pros and cons of efficient productivity; we just went ahead and created a society of consumers who buy more than they need but always less than they want. Now fully acquainted with the problem of waste, we diddle and dawdle about how much we should change our ways. We build faster cars and more complex computers but, as Ellul observes, "the faster the machine, the more serious the accident, the more subtle the machine, the less forgiving the error" (*The Technological Bluff,* 56). For every new telephone enhancement, a new defensive device is needed, and so forth, until we seriously doubt whether progress is being made.

The emergence of genetic engineering will undoubtedly be the next testing ground. Scientists now have the ability to transfer a single gene from any organism—plant, animal, or microbe—into a food crop. Thus, soybeans were modified with a gene from Brazilian nuts to increase protein, but they also cause a strong and potentially fatal reaction in people sensitive to Brazilian nuts. The decision was made not to market the transgenic soybean.

To the extent that we see through this technological bluff, the following general propositions are true. First, science rules as queen only as long as we still believe that every problem has a technological solution. Second, science has bequeathed to us paradoxical paradigms: speed is better, efficiency above all else, producing more is a good in itself, the bottom line is what counts, our resources are unlimited, science will find a way, we have planned for every contingency. Third, every application of technique has a number of positive and negative consequences that are rooted in nature itself, as well as in the method we choose to unlock nature's power.[20] Fourth, and closely associated with the last proposition, we do not so much choose or direct, but are carried by the current of various technological revolutions. Fifth, we will no longer accept the proposition that anything that technique can do is to be permitted. We see the wisdom of limits even though we cannot agree on what those limits should be. Sixth, we are aware that rather than serving the common good, technique will be used for the profit and benefit of a few. Seventh, the belief that science is our last chance to save ourselves is dying, since the self we need saving from is not solely within the domain of an exacting science.

It may be time to reintroduce into our discussion the understanding that humankind always overreaches. Sin is a dimension of self-awareness; as self-awareness deepens, so does the temptation to believe that we can be our own masters. We are beginning to own up to our responsibility for our planet's well being, but not much will change unless we acknowledge that the theological tradition of original sin cannot to be ignored.[21]

Conclusion

The capitulation of science is not unequivocal. Victor Weisskopf, as one of the scientists who thought deeply about how science must change after the atomic bomb, encapsulates several beliefs that motivate science then and now.

> We scientists are optimists. We believe that *rational thought* and planning will be able to rectify the ills that technology has caused. We believe that, at the end, much good will come from the applications of science. But science does

20. I am not arguing that nature is ontologically good or evil. To blame nature for a devastating hurricane is anthropocentric. What does the universe care if there is a human race? It does not. It is in the coupling of ontology and epistemology, where the latter is inherently human, that positive and negative consequences arise. In other words, it is in the "touching of nature" that values are pertinent.

21. Here I am using the term "original" in the sense of "inevitable" without implying the reason, and especially not a genetic reason, why the bent toward self is inevitable.

> not only influence our physical environment, it also *creates our mental environment.* It has deep influences on our thinking and our outlook; it is an integral part of our civilization. (A speech given by Weisskopf during a sit-in at MIT on March 4, 1969, quoted from *In the Shadow,* 26; italics added.)

In the beginning of this book I made the distinction between a view of the world and a worldview, between a paradigm and an episteme. If science has only influenced the way we see the physical world, then it is a paradigm among other paradigms. But if it has become the dominant way we think, the discipline that lays down where and in what sense we may speak of knowledge, then it is an episteme without a rival. I have given reasons, though, why science should not be regarded as the new queen of all human knowledge. Its hallmarks of objectivity and progress have proven illusionary. Hubris has not exactly been replaced, but a taste of suspicion moderates the appetite for yet another bite from the tree of knowledge.[22] Most importantly, in becoming so successful by becoming so reductive, science has backed itself into a corner where it cannot address the perennial questions that are necessary to frame meaning around the larger questions of life and orientation.

How does one reconcile the fact that science has generated a new culture but has fallen from its throne? This question is answered in part by assuming that science's methodological success is one that is shared by the culture as a whole. In the strongest sense, science has provided us with a new way of being in the world. Is it justified, however, to assume that science and modernity are one and the same? The success of science's principal paradigms—principles of mathematics and universal laws, the

22. Some contemporary attacks upon science convey the attitude that science must be humbled. See Bryan Appleyard, *Understanding the Present* (New York: Doubleday, 1992), and to a lesser degree the writings of Mary Midgley, *Evolution as Religion* (Oxford: Oxford University Press, 1988), and *Science as Salvation* (New York: Routledge, 1992). See also John Horgan's national bestseller, *The End of Science: Facing the Limits of Knowledge in the Twilight of the Scientific Age* (New York: Addison-Wesley, 1996).

neutralization or deteleologization of matter, the evolution of all things, natural selection, feedback loops, complementarity, and the belief in progress—have in large measure generated a new culture. These abstract paradigms of thought have become the concrete symbols of how we live—the machine, the city-state, genetic engineering, the atomic bomb.[23] In Foucault's definition of an episteme, an implicit or emergent agreement of what is knowable constitutes a historical a priori, which in a given period "defines the conditions of possibility" concerning what is true. Science introduced a new method of knowing that "won the day" and it became the new episteme—the unconscious, accepted, historical a priori concerning what is real and how the real is known. Nevertheless, its reign of method was intertwined with a host of other cultural shifts: the rise of capitalism, the modern bureaucratic state, totalitarianism, a new mobility, technological advances, and new moral sources. Science has a standing of its own, while the culture is the larger context in which it operates.

At the very core of science's progress is a dichotomy between the promise and the actuality. When the promise fails to materialize, who is blamed? Not science but its application—and behind the application is the human factor. When the theory of evolution was beginning to take hold, it was commonplace to speak of the perfectibility of the human species. That promise is now doubted by almost everyone. If there are holdouts, as I suspect there are, it would be scientists in the field of genetic engineering. Why is it that scientists are more optimistic than the public around them? We do not even have to answer the question; the very observation is worthy of considerable caution. If we are to venture a response, it lies in the narrowness of the narrative story that science permits itself to see. By excising certain metaphysical-ethical considerations, the inclusion or exclusion of perennial questions or the greater health of the environment, for example, the vision

---

23. Gerald Holton makes the same observation. "As a consequence the methods of argument of science, its conceptions and its models, have permeated first the intellectual life of the time, then the tenets and usages of everyday life." See *Einstein*, 43.

of the next technical advance is inevitably myopic. The temptation to overpromise and overreach is compounded when the time between research and application is foreshortened and the boundaries between universities and commercial gain evaporate. This was the point of Michael Crichton's novel *Jurassic Park*. The prominence of the twin disciplines of the history and philosophy of science has mollified the claims scientists make. Even their perspective is too circumscribed, however. The history of science is not the history of humanity, and the philosophy of science is not the wisdom of the ages. As long as science continues to overpromise without regard to our invariable predisposition to overreach, then it surely needs to be confronted by a nonempirical rationality.

Directly or indirectly, the issue is about trust. Just as Galileo's trial was about trust ("Trust me," Galileo said in effect, "I know what I'm talking about"), so Jeremy Rifkin invites us to ask ourselves whom we would trust with the decision of what is a good gene to add to the gene pool and what is a bad gene to eliminate: "Would we trust the federal government? the corporations? the university scientists? a group of our peers?" Rifkin's point is that at one level "we are more than willing to allow the rest of the living kingdom to fall under the shadow of the engineer's scalpel, as long as it produces some concrete utilitarian benefit for us" (*Algeny*, 237). At another level the prospect of human cloning, when it became a real possibility after the cloning of the sheep Dolly, prompted President Clinton to issue an executive ruling barring the use of federal funding to achieve such a technological feat. Where is the standard of measure when we who engineer decide to redirect ourselves? Like the story of Jesus' temptation (Luke 4:1–13) or Goethe's story of *Faust,* a future of unlimited possibilities is indeed enticing. The tree of knowledge filled with delectable delights stands waiting for the extended hand with a voice whispering, "Go ahead and do it, it's your right."

Part Two

The Philosophical and Theological Landscape

CLEARING THE GROUND and setting the scene is a necessary prerequisite to moving forward. The dethroning of theology by a superior method of coming-to-know drastically altered the balance between epistemology and ontology. The philosophical road through Newton led to a new synthesis or fit between epistemology and ontology. Philosophers of the seventeenth and eighteenth centuries not only elevated methodology over ontology, they individualized it (the turn to the subject) and interiorized it (truth is what happens in the mind). Consequently, theology found it necessary to search for an appropriate methodology and this proved to be exceedingly difficult.

The present philosophical landscape is best described as the double crisis in epistemology and ontology. The objective real has been displaced but a softer real and truth claims are about the best argument. In face of this onslaught upon any fit between ontology and epistemology, I argue for a robust ontology and a methodology that includes the question of truth. Jacques Ellul declares that we can only ask the question of truth and attempt to answer it through language. Verbal truths and empirical truths both have their place in the quest for a narrative truth.

In this way we are positioned to critically examine three current models of how science and theology might converse with each other.

Once upon a time there was just one queen and truth was singular and unequivocal. Now in this era of many authorities, none of whom have an all-seeing Eye, the preeminent question is, "How does one adjudicate between rival truth claims?" One possible response is to allow rival perspectives to challenge each other, resulting in a holistic, more encompassing narrative about the kind of creatures we are and the kind of universe we live in.

Chapter 3

Clearing the Ground and Claiming the Real

For the Church there was something Promethean and also Luciferian in the exploits and perception of Galileo. With Galileo final causes became mere superstition, with Galileo the awesome spell cast over humankind by Nature had been broken, and the indomitable reverence of the Greeks for nature replaced by the thirst to plumb the depths of Nature's mystery regardless of the consequence.

—George Bailey, *Galileo's Children*, 50

Did the scientist match his beliefs with what reality is like? No, says Rorty. He made his beliefs conform with what conversational practices had frozen as the background that fixed what everyone would count as real.

—Frank B. Farrell, *Subjectivity, Realism and Postmodernism*, 249

I am certain that since the beginning, human beings have felt a pressing need to frame for themselves something different from the verifiable universe, and we have formed it through language. This universe is what we call truth.

—Jacques Ellul, *The Humiliation of the Word*, 22

– SECTION 6 –
THE DOUBLE CRISIS IN ONTOLOGY AND EPISTEMOLOGY: THE POSTMODERN CONDITION

Introduction

The title of this section presupposes that there is a fit between ontology and epistemology. This, of course, has been a premise throughout *Competing Truths.* It was the basis for defining a worldview in section 1 as the coherent synthesis of theology, philosophy, and natural science, thus making it possible to answer the perennial questions. The modern era shifted the fit in favor of methodology—the real outside the mind became less important than the meaning that was constructed from within the mind. The postmodern condition destroys the fit between epistemology and ontology, constituting the present double crisis. The descriptive strokes here will be understandably broad. An important refinement, though, will be the argument that science and philosophy took different paths in the nineteenth century: science retained a certain confidence in ontology, while philosophy progressively rejected the real as important.

In some form there has been operative in Western philosophy a theory of truth that presupposes a fit between epistemology and ontology. How entities are known must accord with their nature. From a twentieth-century perspective Roy Morrison writes: "Knowledge or even the assignment of 'reality' arises from the continuous *correlation* of empirical factors with the nonempirical facts that emerge in our thinking and in our transactions with the physical universe" (*Science, Theology and the Transcendent Horizon,* 109; italics in original). Prior to modernity, and thus defining premodernity, this fit or correlation was skewed in favor of ontology. Because the cosmological order embodied an "ontic logos" (Greek) or "divine purpose" (Christian), ideas and values were located in the world rather than in persons. The meanings or matterings were somehow inherent in the external world, located behind or beneath the phenomenal world of sensible experience. Aristotle summarily said the "actual knowledge is identical with its object." The privileged locus is ontology, because finding oneself rightly, as opposed to correctly, comes from fitting ourselves to the significance that things already have. Whether it was called the Great Chain of Being, the drama of salvation, or the cosmos as a sacred canopy, the universe was not merely the framework or stage. It constituted a teleological ontology that provided the means for individuals to answer who they were and what role they should play because they knew the nature of what existed.

This teleologically driven ontology was coupled with an epistemology characterized as a natural fitting together. The way in which the world offered itself to be known made it possible for the knower to "read off" eternal truths from the created givenness of the heavens, nature, and history. These truths assumed various forms—ideal types, immutable forms, and teleological purposes, for example. Intuition, deduction, and inspiration moved the subject to grasp what was there but not always apparent. Meditation was given the highest honor, since it was the most receptive form of coming-to-know. Inspiration and intuition were ways of bringing to light what was already present—the opposite of Bacon's method of disrupting nature to see her hidden secrets. Neither the Greek nor the medieval self had difficulty certifying a marriage of the external world and the self.

Modernity began by extending belief in an Aristotelian, ontologically determinate world. Isaac Newton is often situated at the nexus. Was he the last of the premodernists or the first modernist? I believe he represented both. Newton was premodern because of his belief in a teleologically driven ontology and a certain natural way of knowing. The "edging over" to modernity was a matter of his empiricism, which was a mixed bag. Newton was still mesmerized by and pursued a methodology of inspiration and intuition, but at the same time he demonstrated a growing awareness that deduction needed to be supplanted by induction, intuition by experimentation.[1] Newton never doubted the fit between a hard ontology (Newton's hard, individualized, impenetrable, objective, and knowable world) and an empirical methodology. The universe was unequivocally nonself and external. Objects were decisively nonself and external, locatable in a specific space and time as distinct things-in-themselves. Equations and universal laws were unmistakably different from philosophical speculations or theological dogmas precisely because they were anchored in the real world of physical entities that were measurable and predictable.

Once modernity is understood as the triumph of the scientific method, the shift in favor of epistemology is set. Bacon, Hume, Locke, and Galileo realized the limitations of sense perception, common sense, and intuition. This permanently ended any schemes entailing a literal fit between knower and known. Galileo found himself caught in the transition. On the one hand, he fought for the validity of perception and the importance of the real. On the other hand, he realized that experience can be deceiving and valued the importance of verbal arguments of "the better kind" or "with greater justification." What one smells, hears, tastes, and touches could no longer be taken as a bridge between mind and world. The eye needs scientific instrumentation and theory to "truly" see. The consensus was that ecclesiastical doctrines and philosophical deductions had only distorted and muddied the water of perception and representation. The resulting dislocation was, according to Roy Morrison, "from objective realism to subjective idealism with regard to the status and constitution of all the factors that provide order and intelligibility to reality" (*Science, Theology and the Transcendent Horizon,* 58).

---

1. Research has brought to light a Newton who had both a public persona to protect as a scientist and mathematician and a closet persona to conceal as mystic and alchemist. Betty Jo Teeter Dobbs and Margaret C. Jacob explain this incongruity when they write, "Once one grasps the immensity of Newton's goal, many otherwise inexplicable aspects of his career fall into place.... If Newton's purpose was to construct a unified system of God and nature, as indeed it was, then it becomes possible to see all of his various fields of study as potential contributors to his overarching goal." See their *Newton and the Culture of Newtonianism,* 12.

The Fork in the Road

The transition from modernity to postmodernity has nearly reached the point of being overanalyzed, but for the most part the analysis overlooks a substantial rift between philosophy and science. If we understand philosophy, theology, and science as siblings, philosophy did not just mediate disputes between theology and science. Philosophy, as it spread out from Kant, Hume, and Locke through Hegel into positivism and linguistic analysis, left ontology further and further behind. Contrary to the popular picture of science and philosophy as sharing the same bed and begetting a new worldview, I conclude that science did not choose to follow philosophy's turn inward. The impression that science and philosophy have fallen into the same ugly ditch of relativism is an oversimplification, if not just wrong. I wonder whether this distorted historical picture has been filtered through the eyes of historians and philosophers of science.

Our present crisis is largely traceable to Descartes and Kant. In relocating objectivity to the eye of the mind, Descartes broke the medieval fit between epistemology and ontology and opened a Pandora's box. The self and the world were not only wedged apart, the interiorization of truth allowed for a sea of new doubts. Susan Bordo observes that the seeds of postmodern deconstruction were already present in Descartes. In her discussion of Rorty's own assessment of Descartes in his *Philosophy and the Mirror of Nature,* she writes:

> It is via the imagery of an untrustworthy "inner space" that the "mind" is born in the sixteenth and seventeenth centurics. It makes its initial appearance *not* as "mirror," the internal reflection of things "as they are," but as *subjectivity*—the capacity of the knower to bestow false inner projections on the outer world of things. (Bordo, *Flight to Objectivity: Essays on Cartesianism and Culture,* 49; italics in original)

Rorty is correct that a critical relocation had taken place from the Aristotelian and medieval premise of "mind-as-reason" to the new premise of "mind-as-consciousness" (*Philosophy and the Mirror of Nature*, 53). The hope of securing a certain foundation could not help but be endlessly frustrated by all that infects and conditions the self-conscious self, and looming on the horizon were the likes of Feuerbach and Marx, Dostoevski and Freud. The pursuit of pure reason encountered historicism, relativism, pluralism where knowing is inherently contextual, constructed, and pluralistic. John Polkinghorne registers his own consternation by referring to Descartes's "beguiling notion that we can, by seeking clear and certain ideas, construct from thought alone an impregnable metaphysics" (*Reason and Reality,* 6).

Kant believed he had given science a metaphysical foundation, but only because he accepted the "sleeping assumption" (Morrison's phrase) that it was possible to possess a priori categories that were not only universal and necessary but also known with certainty and without distortion. Kant offered his well-known transcendental critique that conceded we cannot know the world as it really is, only as it is experienced as appearance. Consequently, Kant perpetuated a new form of essentialism that was by his own admission meant to counter the skeptical doubt spawned by Hume. Kant's idealism rested on his confidence that there could be a perfect fit between universal or a priori concepts and the world as it is experienced but not as it is. Robert D'Amico explains Kant as hoping "to control the destructive impact of this admission (things-in-themselves are inaccessible) by showing that conditions for the possibility of the world as appearance are universal, a priori (prior to experience) and humanly invariant" (*Historicism and Knowledge,* 132).

Kant's transcendental idealism had at least two deleterious implications for modernity. It wrecked ontology by accepting the presupposition that there are no "things-in-themselves" that exist independently of the human mind, that function universally, and that have their own identifiable intrinsic features. Secondly, it started modern thinkers down the road to believing in a singular and certain methodology capable of finding universal or a priori conditions for our experience of the world. Kant wanted to save epistemology by staying out of the muddy water of experience, and this became the weakness of modernity. Ever since Cartesian interiorization of truth, philosophy has inched its way toward acknowledgment that experience is not merely the sensual portal for coming-to-know. Experience is shaped by the language-in-use of the knower or observer; therefore, all knowing is conditional.

The resulting interiorization of truth and its decontextualization meant the separation of self from world, knower from known. This became the premise for empiricism with the unfortunate side effect of positing two forms of truth that were declared inherently inimical: truths of the mind and truths of the physical world.[2] For all their efforts to the contrary, Descartes and Kant dragged epistemology into a new world of suspicion, because mind as self-consciousness meant that one was continually forced to purify truth of its connection to the body and its cultural-historical conditioning. Truth needed to be decontextualized until nothing was left but clear and distinctive ideas. The resulting outcome was the inevitable reduction and fragmentation and eventual demise of truth as word.

Descartes wanted to decontextualize the process of knowing, a proposition that postmodern philosophy rightly rejects as completely wrongheaded. One would have expected empiricists and rationalists to forge a new marriage founded upon objectivity. And they did so for awhile.[3] They were joined by their common rejection of a teleologically driven ontology and their desire to

2. Anders Nygren, *Meaning and Method*, 80–101. Here Nygren discusses the reduction of one kind of truth to the other, judging one kind of truth by the other, and the limitations of both kinds of truth.

3. Charles Coulston Gillispie's *The Edge of Objectivity* is a book about the persistent confidence in the ability of scientific methodology to progress until objectivity encompassed all rational endeavors. I see this flight to objectivity as science's way to protect itself from the subjectivity inherent in all Cartesian-type philosophy. Historically, objectivity was predominant during the transition of dethroning theology and the ascendancy of empiricism.

distance themselves from speculation, opinion, and external authorities. The seeds for divorce, however, were present from the start. Empiricists would insist that all knowledge has its source in experience. They could not abide the proposition that sense experience should be made inconsequential; they were committed to the opposite conclusion that truth could be decided only when tested against sense experience, with the help of instruments. Rationalists, on the other hand, wanted to be purists, which translated into the time-honored tradition of idealism and some form of essentialism. Rationalists were forever pursing ideal types, immutable forms, created species, inerrant texts.[4]

A defining moment for science was Niels Bohr's conclusion that the universe could not be known as it really is. Philosophers had already reached the same conclusion but it was a different story for scientists. Science did not relinquish its confidence in the givenness of the external world as verified by mathematics, experimentation, and a progressive march toward theories that worked. Bohr's defining moment, however, would not go away. The problem was deeper than the time-honored distinction between things in their essential nature and knowing them through the lens of experience. A universe constituted by particles that are themselves defined and understood by their relationship to other properties requires a language that is more relational, metaphorical, colorful, and metaphysical. Science did not so much mimic current philosophical trends—it began to do its own philosophy in the hands of such titans as Bohr, Heisenberg, Russell, Whitehead, and Einstein. Scientists became more and more confident of their own methodology. The philosophical turn to the subject was incongruent with the truth-model known as correspondence—the fitting together of language and world, the thing-in-itself and its representation. This model was a cornerstone of empiricism because it secured a sense of objectivity. It became more and more suspect as more and more subjectivity was interjected between

4. See David Hull's excellent chapter, "Essences," in his *Darwin and His Critics*.

the two poles. The argument for objectivity had little value for philosophers, because it had been discredited by epistemology. This was not true for scientists, who continued to use it as a working model. When it did begin to falter, it did so on the ontological axis. As the ontology science discovered became softer—the inclusion of indeterminacy, field theories, probability, chaos, parallel universes—science had to make space for a plurality of interpretations, the idea of complementary truth, the historical-cultural nature of all knowing.

Is the historical picture clarified when theology is included? Although this question will be discussed in the next section, it is interesting to ask here if theology favored one sibling over the other. When the epistemological-ontological fit of a Newtonian universe began to unravel, theology was in a methodological quandary. Therein lies the clarity. In the realm of generalizations, theology was swallowed up in the shift toward epistemology that marks modernity. The right method became its preoccupation. While Jeffrey Stout is writing about authority, the following statement clearly shows theology's initial tilt in favor of the interiorization of truth.

> For Descartes, as for Luther before him, what most matters in life is no longer played out in the dimensions of community and tradition. One discovers truth in the privacy of subjective illumination, and this truth is underlined by a kind of self-certifying certainty. Community, tradition, authority: these have all started to give way to the individual, his inwardness, his autonomy.[5]

The problem became this: philosophy in the hands of Descartes did not become the dominant sibling. Science, with its promise of objectivity and progress, won the day and became the predominant cultural mentality. The question then becomes why theology

5. Jeffrey Stout, *The Flight from Authority*, 49–50. Stout would say that it was an epistemological shift from the age of authority to the age of probable reasoning.

proved to be so decidedly unsuccessful at forging an empirical-like methodology. I believe the answer is twofold. First, theology has an epistemological-ontological fit that is decidedly different from science. Second, theology has always had an affinity for philosophy. Why? Because theology by tradition is a discipline about the truth of words or word-truth. Theology is metaphysical by nature since it invariably deals with the deepest kinds of questions we can ask. Thus, theology's desire to claim for itself an empirical methodology was inherently opposed to what Charles Taylor calls the "constructive power of language" (*Sources of the Self*, 198).[6] The result was that in the twentieth century theology twisted and turned to find an appropriate methodology (see next section).

Premodern to Modern to Postmodern

The premodern era was indubitably a teleological constituted ontology. The perennial questions about human nature and the nature of the cosmos were addressed by looking into the givenness of the world. As a species "set apart" (created in the image of God), humankind knew its rightful place by proper observation (the Greek tradition) and by meditation (the Christian tradition). The modern era is best characterized as a hard ontology wedded to objectivity but not without marking the parallel movement to seek truth within the interior of the mind. Charles Taylor succinctly states one polarity of modernity: "Knowledge comes not from connecting the mind to the order of things we find but framing a representation of reality according to the right cannons" (*Sources*, 197). This was the tilt toward epistemology where language organizes experience. The other modern polarity, empiricism, understood words to get their meaning from things

6. Taylor brilliantly unpacks the road to the interiorization of truth, especially in his discussion of Descartes's disengaged reason and Locke's punctual self (*Sources*, chaps. 8 and 9). Taylor's phrase, the "constructive power of language," has a theological ring to it and becomes food for further thought as I consider what voice theology should reclaim in part three.

in the world (or states of affairs) they represent. Experience and all that experience brings with it—culture, biases, suppositions—were not to be denied but they were not to be counted as a decisive factor because the scientific method retains procedures to verify and reach consensus. The postmodern condition, on the other hand, is the complete disruption of any fit. Epistemology became a separate discipline fiddling with language analysis and concluding that words get their meaning as they reference other words.[7] Philosophers and historians of science questioned why empiricism should be "the last bastion of epistemic privilege," for its history belies a record of "self-recruiting, normatively defined, variously directed, and often sharply competitive scientific communities."[8]

The crisis would be deep enough if it were merely a separation of facts of the world from truths of the mind. In addition, postmodernity requires that both fact and truth be placed within quotation marks. Facts are no longer factual and truths are no longer truthful. The former have become "fictions embedded with theories," and the latter have become "the better argument or explanation." Since meaning can no longer be located in the nature of the thing itself, it must be contextual and intratextual. In short, neither epistemology nor ontology is capable of truth unless it is "facts" or "truth." Finally, postmodernists and philosophers of science have lost confidence that the world confers meaning. The scale has been so tilted to the side of epistemology that the world is not sufficiently other to have its own determinacy.

Reconstructing the Fit

In his study of the Enlightenment, Colin Gunton highlights the false choice that broke apart the premodern fit between an epis-

7. Jeffrey Stout adds a further clarification: "Descartes made epistemology the basic level of inquiry for generations of philosophers, but Frege made the theory of meaning seem more basic still" (*Flight from Authority,* 77). Once this gate was open, the analysis of language would become paramount and all sorts of intellectuals, but mainly theologians and philosophers, could enter the conversation since meaning is anyone's game.

8. I am quoting Clifford Geertz in his assessment of the legacy of Thomas Kuhn. See Geertz, *Available Light* (Princeton, N.J.: Princeton University Press, 2000), 163.

temology of intuition and inspiration and an ontology with its own divinely endowed significance.

> The Enlightenment alienated the meaning of words from the meaning of the world. Because the world was denied any inherent meaningfulness, the locus of meaning came to be found in the human mind. . . . The whole error at the root of the enlightenment tradition has been its presentation of a false dilemma between a transparent rationality and one that is imposed by the human mind. (*Enlightenment and Alienation*, 144–45)

Aristotle and Plato were able to envision a transparent rationality to match a transparent cosmos, but Bacon, Descartes, Hume, and Locke could not. The model of correspondence would die the death of a thousand qualifications. It would survive just as long as objectivity remained a valid possibility. It remained a working hypothesis for scientists far longer than it did for philosophers, because they spent their days peering through telescopes and microscopes and measuring observable things. We need to remember that those entombing the correspondence model were the philosophers, because certainty was being relocated from the world as-it-is-out-there to what our minds present to us immediately and directly. The postmodern insight was to see the flaw in this epistemic assumption concerning immediacy. What the mind "thinks" is invariably provided to it by experience as it has been interpreted by language. But language also has a second function besides interpreting what is external to the mind. Northrop Frye more accurately delineates the two primary functions of language when he writes: "Here we have two structures, the structure of what is being described and the structure of the words describing it" (*Words with Power,* 4). The model deemed irrelevant is the Newtonian empirical model where language is meant to be a literal or exact fit. Scientists clearly do make softer claims about reality and how it can be described, but it does not mean they no longer see themselves as fitting together words, sentences, and symbols with a universe that is discovered more than it is created. The obstacle for many, including George Lindbeck and Nancey Murphy, is to presume that if there is to be some basis for a theological realism, it must be with this defeated Enlightenment-romantic understanding of correspondence. When the choice is between either a direct correspondence between signifier and signified or no relationship at all between words and world, we are being offered a false alternative. Of course there is a space or separation between experience and expression and between experiment and theory formulation that necessitates the inescapable mediation of human language. There is not cause, however, to disparage the real as important. I am not arguing for the reestablishment of a certain foundation based upon an empirical model of correspondence or the objectivity of language. Language is fallible and context driven, and reality is not a "thing-in-itself" to be grasped and captured by language. The model I present is much more dynamic because language is elastic and reality is ambiguous.

Conclusion

In the best of all possible worlds, there would be a natural and balanced fit between epistemology and ontology. Thomas Torrance speaks about this ideal fit when he writes: "You know something only in accordance with its nature, and you develop your knowledge of it as you allow its nature to prescribe for you the mode of rationality appropriate to it" (*God and Rationality*, 52). We have nevertheless distanced ourselves from this Enlightenment belief in an ideal fit and understand it, rightly so, as naïve, but are we therefore justified in denigrating both language and object to the point that we dismiss either or both? The denigration has taken place on at least two fronts: by squeezing the givenness out of reality because it cannot be objectively given and by treating the truth of language as no truth at all because it is invariably historical and relative. The contrary thesis of this book is that truth can be constructed from the subjective side as verbal and discovered from the objective side as empirical. Torrance is fully justified in his insistence that how we know and what we know are joined, and if "we are to know some object in accordance with its nature, it is that same nature that must prescribe the mode of its verification."[9] How we come to know the nature of a desk is different from how we come to know another person, and the nature of a star and the nature of God require methodologies that ask different questions and have different ends in mind. John Polkinghorne reminds us that we know the everyday world in its Newtonian clarity and we also know the quantum world in its Heisenbergian uncertainty. Our knowledge of entities, he concludes, "must be allowed to conform to the way in which they actually can be known" (*Faith, Science and Understanding,* 7). The proper conclusion is to profess that we are no longer beguiled by the Enlightenment hope for a universal epistemology, while not submitting to the pessimism of truth as little more than local fictions. Since the world is neither mute nor transparent, there must be a mutual learning dialogue between self and the world. Of the three perennial questions, the third one invites us to understand the universe and our place within it holistically, but in one important respect we are impeded from stepping onto this path. We have done no less than build barriers between what is true and what is real, because we have little tolerance for different ways to inquire after what is true about what is real. And while there may be particular fits between epistemology and ontology that are either more appropriately verbal or empirical, any truth that matters—and truths that matter inescapably address the perennial questions—tries to be sensitive to the richness of the poetic and historical and the complexity of the

9. Thomas F. Torrance, *God and Rationality* (London: Oxford University Press, 1971), 52. Torrance adds a theological kicker: "He [Jesus Christ] alone can put to us the true questions that make us free for truth" (54). Is this a theological boundary that science cannot cross? Because theology and science are distinct disciplines, it would seem so, although there are individuals who let Christ ask the most important questions about who we really are without letting that faith disposition cloud their empiricism.

mathematical and universal. Herein we contemplate "who and what is man within the cosmos" (Karl Barth), and to this purpose completing truths may be advantageous as well as necessary.

The world is not merely a stage upon which the actors play their parts but the context that shapes the actors. The drama is more than the roles we play out; the script incorporates the pattern of events that arise independent of ourselves. The evolution of the universe does write its own drama, a drama that is nevertheless incomplete without the reflection and understanding that is imparted by the human mind. Truth is to be found and constructed by looking at the eruption of decisive significant events, by the protrusion of the memorable, and by the narratives that unify past and future trajectories spiraling forth from those events.

– SECTION 7 –
THEOLOGY IN SEARCH OF A METHODOLOGY

What was theology to do? The ontological view of the world it knew and defended with theological truths was slowly and steadily being transformed. Its worldview of perfect circles and fixed planets, of created species and ladders of beings and meaning, had been turned upside down and historicized. Its beloved perennial questions were co-opted by secular interests and dictators. Most important, theology did not have a methodological leg to stand on. The triumph of the scientific method bent theology like a huge black hole. Theology was sorely in need of new foundations because the former ones of verbal truths, inerrant texts, miracles, and acts of God had suffered heavy casualties, some of them fatal. The Platonic-Christian eye of the soul that saw behind appearance to the "real thing" had no place in a scientific age that no longer believed in the soul or sin or inner significance or innate properties. The sovereign-like aura of science had effectively removed religious truth from the domain of matters of fact. Religious knowing was suspect, isolated, and considered irrelevant.

Any student of theology soon becomes confused and dazed tracing the courses theology chose to take. Would theology follow the lead of science? Not much of anything "out there" as reality could be considered religiously meaningful. Science did entice theology not so much by its ontological confidence as by its epistemological success. Was theology to follow the lead of philosophy? Philosophy had turned analytical but with a negative, disdainful tone. The more theology looked to philosophy for metaphysical underpinnings, the more it discovered that philosophy had killed metaphysical idealism as well as objectivity. Like philosophy, theology retreated to interiority and became mired in the same quicksand of language, self-consciousness, history, relativism, and subjectivism. It would have been different had science and philosophy provided a unified framework, but science and philosophy had gone their separate ways. The crisis in epistemology and ontology, that thinning out and softening of truth claims, meant that theology would either have to harden old positions, imitate science, or forge its own identity. In the late nineteenth and twentieth centuries, theology did all three. Above all else, theology became obsessed with methodological concerns.

Theology Defends Her Foundations

The threat was not modernity as such but secularism, and likewise, not science per se but an atheistic science. Conservatives, in contrast to liberals, were not willing to concede certain intellectual property, which explains why there were confrontations on specific issues, such as creation. They reacted strongly to any threat to erode Christian foundations. In face of that threat there was a sustained effort to reassert theology's authority. This happened on various fronts, including the inerrancy of Scripture, the propositional nature of revelation, and the confessional core of systematic theology. What also characterized conservative theology was the unwillingness to separate theological truths, such as the divine origin of human life, from the facts, scientific and historical. Like the Galileos and the Newtons, it mattered that one's science not be divorced from one's Christian beliefs. Galileo and especially Newton were motivated by the unity of knowledge, while conservatives were motivated by a literal reading of Scripture. Contrary to popular perception, evangelicals and fundamentalists were enamored with science, for they saw in this discipline traits and principles to be admired; namely, rationality, facts, and the correspondence model of truth. The use of the word "fact," for instance, litters the language of conservatives. Charles Hodge, who was a pivotal figure in the hardening of theology's position, writes, "The duty of the Christian theologian is to ascertain, collect, and combine all the facts which God has revealed in the Bible. . . . "[10] Thus, the supposedly antiscience posture of conservatives and fundamentalists was a reaction to a science that had become philosophically married to determinism and materialism. Because conservative theology took the model of correspondence seriously, it also took potential points of conflict seriously.

There is a certain irony that conservative theology could be antiscience but still side with the scientific correspondence model of truth. Conservative theology was repulsed by the notion that words are true because they refer to other words; it correctly sensed that philosophy was headed toward relativity. In her book *Beyond Liberalism and Fundamentalism,* Nancey Murphy brings a fresh understanding to how conservatives (fundamentalists) differ from liberals. She distinguishes between an "outside-in" (conservative) and an "inside-out" (liberal) philosophy. An outside-in approach to epistemology is where "words get their meaning from the things in the world to which they *refer,* or

10. Charles Hodge, *Systematic Theology,* 3 vols. (New York: Scribner's Son, 1891), 1:11. See also the language of a more contemporary conservative theologian, Augustus H. Strong, *Systematic Theology* (Philadelphia: Judson, 1949) in Nancey Murphy, *Anglo-American Postmodernity* (Boulder, Colo.: Westview, 1997), 90.

sentences get their meaning from the facts of states of affairs they *represent*" (italics in original).[11] The operative belief is that facts of nature, or history, or the universe are revealed or present themselves to our faculties. Language, then, is not about the meaning of words but about which words best or faithfully describe the way the world is and the way God is. Using the right words is important for both theologians and scientists insofar as they desire to describe a fit between language and the real.[12] Murphy defines an inside-out methodology as beginning with the contents of the mind "and then seeks to give an account of the world in those terms" (*Beyond Liberalism,* 28). This is the Cartesian turn to the self that became manifested in modern philosophy and philosophers from Schleiermacher to Bultmann, from existentialism to the "experiential-expressive" theologies. It is easy to understand why conservative theology was so averse to the subjective nature of what Murphy calls experiential foundationalism. Both on the ontological side and on the epistemological side, conservative theology fought tooth and nail to preserve what is external to us. Inspiration must be wedded to the external authority of Scripture and revelation must be God's coming to us.

The Newtonian synthesis was a solid foundation built upon the correlation of physical reality with mental concepts. The Newtonian worldview had a place for a benevolent God who could and did intervene when necessary. The world had a certain fixity and permanence that was easily hierarchialized and populated with fixed species. There was an accuracy in what is asserted, a preciseness in words, a reasonableness about propositions. In

---

11. Nancey Murphy, *Beyond Liberalism and Fundamentalism* (Valley Forge, Pa.: Trinity Press International, 1996), 38.

12. There is a third understanding of language to be inserted between conservatism and liberalism—the realism of Karl Barth. See the insightful essay by George Hunsinger, "Beyond Liberalism and Expressivism: Karl Barth's Hermeneutical Realism," in *Disruptive Grace: Studies in the Theology of Karl Barth* (Grand Rapids, Mich.: Eerdmans, 2000); 210–25. I have difficulty thinking of Barth as a realist, just as I do Hans Frei (see below). Hunsinger is right on target to identify Barth's understanding of language as having the possibility of imparting authentic knowledge about God where language about God, while it may be analogical, is analogically *reliable*. Thus the right words do matter because only the right words bring to expression the reality of who God is.

this marriage of epistemology and ontology, science and theology shared an important understanding of how language functions, namely, as a bridge that ties together universal concepts and empirical events. The subject matter, of course, was different, so that clear and distinct ideas became inspired words and propositions; empirical events became the historical acts of God.

In a similar vein, Nancey Murphy calls attention to the very different theories of language operative in conservatism and liberalism. It is a distinction that I have raised previously and will elaborate on here. Where ontology predominates and literal truths are possible, language can move from outside-in resulting in propositional knowledge. In this scenario, language's function is to make the connection and bring to expression what is discovered or experienced. Meaning is rooted in the ontology or givenness of what is there. This theory is what David Tracy calls romantic because it believes language articulates a fact rather than an interpretation, and what I call the communication model because language communicates what is prelinguistic.[13] Where epistemology predominates (inside-out), language has a much stronger role because it shapes or interprets the experience. Language has the power to construct; because it has this power, meaning is not some real thing that exists prior to experience—language arises from the languaging of the experience. This distinction of how language functions is one of most fundamental differences between conservatives and liberals and it has not received the attention it deserves.

Murphy describes both conservative and liberal theology as foundational: conservatism with its foundation of a literal interpretation of Scripture and revelation as propositional, and liberalism as foundational in the kind of experience that is available to all people (*Beyond Liberalism,* 37, 50). As most read the history of liberalism since Schleiermacher, it has been anything but foundational in the same sense as conservative theology. At

13. David Tracy, *Plurality and Ambiguity,* 48–49, and the discussion in section 5, pp. 98–99.

the very least one must make a distinction between modern and postmodern liberal theology, but even with this distinction it is a stretch to consider twentieth-century liberal theology as foundational. If there is an argument to be made in this regard it would be that liberal theology tried to become foundational in an epistemological sense; that is, it appealed to the universality of experience and universal experiences. Thus Barth's assertion that modern theology begins with Schleiermacher

The strength of conservative theology is its determination to resist the temptation to allow modern secularism to dictate how to do theology. Conservatism, however, did not escape the gravitational pull of science's success, but in this regard it admires those very traits that have virtually disappeared from science. Conservative theology and science made good partners insofar as they both used a propositional-referential epistemology and a hard ontology. Science has steadily moved away from any naïve kind of fit, and so have many conservative theologians who eschew literalism but embrace narrative theology. The dilemma, as Jeffrey Stout perceptively illuminates it, is how theology can retain its integrity without undo reliance upon self-authenticating criteria. Conservative theology continues to make its stand by beginning with an external, self-authenticating authority, namely Scripture. With the generation of evangelical scholars including Carl F. H. Henry and Clark Pinnock in America and G. C. Berkouwer and Helmut Thielicke in Europe, there was a new sensitivity to the charge of being irrational. These "new evangelicals," as they became known as in the 1960s and 1970s, felt obligated to correct this impression and to make conservative theology distinguishable from fundamentalism.[14] More recently, a second generation of new evangelical theologians, familiar with and sympathetic to the postmodern critique of foundationalism, is grappling with whether or not evangelicals can be nonfoundationalists, a new

14. See Richard Quebedeaux, *The Young Evangelicals* (New York: Harper & Row, 1965) and R. Coleman, *Issues of Theological Conflict,* chap 1 ("The Forging of a New Middle").

account of biblical authority, and the issue of realism.[15] The future of conservative theology will be determined by its search for an epistemology that can make inspiration and revelation essential without depending upon a model of correspondence that is literal and propositional.

Theology That Becomes Scientific

Process theology[16] is the clearest example of the way that theology imitated science; it may be the only modern form of theology to take the ontology of science seriously. Process theology stands apart from other theologies because it operates out of a philosophical framework and, to be more specific, out of the philosophy of a particular scientist, Alfred Whitehead. Charles Hartshorne was a major interpreter of Whitehead, while a number of theologians, such as Schubert Ogden, W. Norman Pittenger, John Cobb, Daniel Day Williams, and Pierre Teilhard de Chardin, pursued further Christian adaptations. Whitehead was interested in developing a comprehensive framework that would accommodate both science and theology. Two of his more enduring contributions were making a place for "event" from Newton's worldview and "evolution" from Darwin's view of the world. An event, whether physical or historical, exhibits both the freedom of alternative possibilities, the event's own self-creation, and the gathering up of the past while moving into the future. Time moves forward, then, in an evolutionary fashion by exercising one chance alternative over another, but each "decision" is conditioned or limited by its past decisions and in a paradoxical way is lured ahead by God. Whitehead would say that God is in the process and the process is in God; that is, God is not absent from any event, nor can any event be attributed solely to

15. One of the best examples of the new postliberal evangelicals is *The Nature of Confession,* ed. Timothy R. Phillips and Dennis L. Okholm (Downers Grove, Ill.: InterVarsity Press, 1996).

16. See Ian G. Barbour for a summary and critique of process theology in his *Religion in an Age of Science* (San Francisco: Harper & Row, 1990), 230–42.

God. Order and novelty arise from God's offering of alternative possibilities. All this happens within the evolution of all things.

Whitehead and Tillich pioneered the modern thrust of reconfiguring theology away from classical theism. In *Science, Theology and the Transcendental Horizon*, Roy Morrison comments: "Tillich spent approximately fifty years of his career (about 1915 to 1965) philosophically removing every conceptual and methodological trace of classical theism that he encountered" (180).[17] The reorientation moved the center of theology from being to becoming, from eternity to temporality, from necessity to contingency, from self-sufficiency to relatedness. One can recognize a parallel movement in science and theology. Just as science began to question an ontology of things-in-themselves, so theology followed by beginning to question an ontology where God is a being-unto-himself. The latter was the classical assumption that the impassivity of God was predicated on the theological premise that to be affected by something outside oneself is an imperfection. More than a few theologians believe Whitehead and Tillich sacrificed the personal God of Scripture in the move toward being scientific, but very few have not been influenced by their deep appreciation of a modern scientific mindset.

This mindset is best characterized by the adoption of an ontology of becoming. Unlike conservative theology, with its love affair with a Newtonian worldview, liberal theology gradually but thoroughly incorporated Darwin and Einstein. A hard theology with hard truth claims gave way to softer ones. Reality was not identified with that-which-is but with all that-is-becoming. Here is the beginning of the crucial shift to an understanding of God who is influenced by the world, what Whitehead called "God's

17. Paul Tillich has a Roman Catholic counterpart in Leslie Dewart, because he conducted the most thorough de-hellenization of dogma. Dewart, *The Future of Belief: Theism in a World Come of Age* (New York: Herder & Herder, 1966) and *The Foundations of Belief* (New York: Herder & Herder, 1969). Dewart took with utmost seriousness the historical nature of coming-to-know and its result: "Truth is not the adequacy of our representative operations, but the adequacy of our conscious existence. More precisely, it is the fidelity of consciousness to being" (*Future*, 92).

consequent nature" and Hartshorne described as "God the sympathetic participant."[18] It is easy to underestimate the influence of a more relational model of reality, which still marks contemporary theology.[19] The effect is a gentler God who persuades and lures creation rather than one who controls and intervenes, one who works patiently and unobtrusively.

The Turn to Methodology

The movement toward hermeneutics represents in its purest form the search for a proper methodology. Stated succinctly, theology became self-conscious that it needed an acceptable method of coming-to-know in order to be received back into the public debate about matters of import. Theology was certainly not then, and still is not, cognizant of how much it reacted defensively to the reign of scientific method by trying to become scientific. Theology has tried one oar and then another: from turning anthropocentric (Schleiermacher) to dogmatic (Karl Barth), from demythologizing religious truth claims (Bultmann) to insisting upon the factual nature of religious belief (fundamentalism), from integrating a philosophy of science (process theology) to broadening its agenda to include the oppressed and nature (liberation theologies and eco-theologies), and finally, simply to bracketing out questions of truth (existentialists and deconstructionists). It is not unfair to say that methodological defensiveness was the chief characteristic of theology after World War II. In the 1950s, 1960s, and 1970s some of the most gifted theologians were doing hermeneutics in place of systematic theology, because the former was seen as a necessary step before the latter could be taken up again.

---

18. Leslie Dewart writes: "The preoccupation with God's existence which characterized post-patristic thought, and thence post-medieval philosophy, is a result of conceiving God as an actual object of thought" (*The Future of Belief*, 185). Dewart chose to put the emphasis on God's presence and absence manifested in and through that-which-is.

19. Not so understandable and also disappointing is the way scientists such as Paul Davies seem stuck on a classical depiction of God and ignorant of process theology and other theologies sensitive to a more relational God.

The initial historical impetus was the debate between Barth, Bultmann, and Brunner, which focused attention on such issues as a proper starting point, the legitimacy of natural theology, the role of philosophical presuppositions, how to do historical studies, and the hermeneutical task of translating and interpreting.[20] As hermeneutics blossomed, it became a catch basin for nearly everything—the quest for the historical Jesus, biblical theology, the bridging of the New and Old Testaments, and historical studies in general. The heartbeat of each turn was always the need to justify one's theology. Even into our times, the need to begin with methodology is paramount. John Dominic Crossan's *The Historical Jesus* (1991) is a prime example of how this need to be methodologically correct continues.[21] It was deemed necessary to clarify one's methodology everywhere and in all endeavors, because everyone believed that no theology could succeed or be justified without the right methodology.

Theology Reclaims Its Self-Identity

In the midst of the concern to be methodologically correct is the passion on the part of a few theologians to be faithful to the core beliefs of Christianity. Characteristic is a Kierkegaardian-Barthian unwillingness to argue with modernity because conversation with science invites compromises in order to be an accommodating participant. The major agenda became how to restore the epistemological and ontological priority that belongs to God, a priority that is understood as lost when engagement with the modern conceptual world is given a privileged place. Among the contemporary voices, I would include Jacques El-

20. See James D. Smart, *The Divided Mind of Modern Theology: Karl Barth and Rudolf Bultmann, 1908–1933* (Philadelphia: Westminster, 1967); James M. Robinson, ed., *The Beginnings of Dialectic Theology* (Nashville: John Knox, 1968); Gareth Jones, *Critical Theology: Questions of Truth and Method* (New York: Paragon House, 1995).

21. Crossan is typical in making methodological consideration paramount. His lengthy appendices and finding a historical (objective) core is a sharp contrast to Bultmann's declaration that the actual words of Jesus are inaccessible and theologically inappropriate. See his *The Historical Jesus* (New York: HarperSanFrancisco, 1991).

lul, Stanley Hauerwas, Howard Yoder, Ronald Thiemann, Hans Frei, George Lindbeck, Garrett Green, Bruce D. Marshall, George Hunsinger, and William C. Placher.

Because philosopher Jeffrey Stout represents an outsider's perspective, he sees with unusual clarity the double bind that Christian theologians face. In *The Flight from Authority* (1981), Stout argues that since Hume and Kant, Protestantism has either recommitted itself to maintain what is distinctive and authoritative or has sought a common ground by a process of accommodation.[22] The full force of Stout's analysis is only felt when the Christian community remembers its faith is about paradox and mystery and not about a better argument. Thus, Stout can write with a degree of sarcasm:

> MacIntyre is therefore right to say that "the lack of confrontation is due not only to the directions in which secular knowledge is advancing but to the direction in which theism is retreating."
>
> Theology since Barth is a sad story. . . . Those theologians who continue to seek a way between the horns, and thus to remain within the secular academy without abandoning the community of faith, have often been reduced to seemingly endless methodological foreplay. (*The Flight,* 147)

The challenge for theology as a discipline is to retain its distinctiveness without becoming irrelevant. Stout suggests several ways this might be done, but none is more poignant than his statement, "Christianity, if it is to be true to itself, will have to make ontological claims" (*Flight,* 143).

In some ways, this represents the classic conservative-liberal choice. Hans Frei, George Lindbeck, and the Yale school of theology have gained prominence by presenting an alternative that

---

22. Stout, *The Flight from Authority,* 97, also 129, 146; Stout uses as a primary source MacIntyre's lectures entitled "The Fate of Theism," which were published in *The Religious Significance of Atheism* (New York: Columbia University Press, 1969).

is neither conservative nor liberal.[23] On the one hand, they are not foundationalists (conservative), because they acknowledge the historical character of coming-to-know and reject a romantic understanding of language where language comes after the fact of discovery and cognition. Nor do they have much use for a model of language that brings together an external reality with its verbal representation. In a well-known watchword they announce, "the biblical word absorbs all other worlds" by means of an intratextual realistic interpretation of Scripture. Lindbeck adds further clarification by emphasizing a hermeneutical method that is inside-out—but the inside is not the self-authenticating external authority of Scripture but the practiced faith shaped by the Scriptures.

> Intratextual or text-immanent meaning is constituted by the text, not by something outside of it such as Marxist, Freudian or Nietzschean causal or genealogical explanations in terms of economic, psychological or power factors. Nor is it only extratextual explanatory frameworks such as I have just listed which are irrelevant to determining the sense of the text. Also excluded are historically reconstructed referents, (human) authorial intention or extratextually known ideas presumed to be present in the text allegorically or otherwise.[24]

Ronald F. Thiemann also appreciates the epistemological dilemma encountered by this third way.[25] His historical analysis uncovers two facts. The first is that modern or liberal doctrines

---

23. By no means homogenous, the Yale tradition of defying categorization begins with H. Richard Niebuhr and continues through George Lindbeck and Hans Frei and a younger group of scholars trained at Yale—Garrett Green, Stanley Hauerwas, Bruce Marshall, William Placher, and Ronald Thiemann. Placher's *Unapologetic Theology* (Louisville: Westminster/John Knox, 1989) is very representative of this "middle course" theology. See also Gary Dorrien, "A Third Way in Theology?" *Christian Century* (July 4–11, 2001): 16–21.

24. George Lindbeck, "Atonement and the Hermeneutics of Intratextual Social Embodiment," in *The Nature of Confession*, 227.

25. See n. 23. This third way is the Barthian-Yale tradition.

of revelation inevitably became epistemological doctrines. Second, the epistemologies of three representative liberal thinkers—Locke, Schleiermacher, and Torrance—depended upon "a notion which grants to God ultimate causal responsibility for our knowledge of him."[26] Thiemann has in mind the epistemological mode of direct, unmediated revelation. He rejects the model as a relapse into a foundationalism that he defines as the belief that "knowledge is grounded in a set of non-inferential, self-evident beliefs which, because of their intelligibility, is not constituted by a relationship with other beliefs" (*Revelation,* 158, n. 20). God's self-disclosure, then, is not found in an Archimedean point, such as timeless truths, a revelatory essence, textual factualness, or normative tradition, but in the prevenience of divine promise (that is, God's gracious coming to us). The foundation for theology, then, is "the continuing reality of God's active presence among his people" that is not seen nor fully experienced and must be expressed by a confession of faith (*Revelation,* 80).

Thiemann believes the critical move is a reorientation of epistemology that restores God as the gracious initiator and fulfiller of God's own promises. In good Barthian fashion, faith's coming-to-know requires the simple acknowledgment of God's gracious revelation when and where it occurs.[27] Thiemann is correct in sensing that the collapse of epistemological foundationalism means we cannot justify our truth claims by noninferential, self-evident data. I am not convinced, though, this third way has restored an epistemology authentic to the Christian faith, because it has not been fitted with an ontology that is more than intratextual or introverted.

---

26. Ronald F. Thiemann, *Revelation and Theology* (Notre Dame, Ind.: University of Notre Dame Press, 1985), 43.

27. Compare this radical and uncompromising Barthian statement by William Placher: "If God is really transcendent, then there is no epistemological path from us to God, and everything we know about God comes at God's initiative." *The Domestication of Transcendence* (Louisville: Westminster John Knox, 1996), 186.

Hesitant Move toward Realism

One can understand the turn to experience as an attempt to restore a viable ontology coupled with an acceptable epistemology. The often-used phrase, "turn to the self," does not explain enough. The turn to the self is usually viewed as an attempt to reinstate an acceptable starting point for religious statements to be heard by the culturally disaffected. Friedrich Schleiermacher's starting point was to show that Christian beliefs are grounded in a deeper universal form of human self-consciousness than one grounded in Christian doctrine. Christian revelation was repositioned as ultimately validated by an intuitive, universal religious experience. By the early twentieth century, according to James D. Smart,

> This emphasis upon experience, combined with the contemporary enthusiasm for the methods of natural science, seemed to offer theology a new and *objective* basis. In religious experience it would have a body of data upon which it could go to work in a thoroughly *scientific* manner and thus gain for its conclusions an objective validity that no *reasonable* man would be able to resist, any more than he could resist the findings of other sciences. (emphasis added)[28]

At work behind the scenes is the powerful influence of empiricism as a method of induction. It was also a search for a proper ontology, the real life experiences of human beings.

It might have worked except that language does not just bring to expression clear and distinct ideas or revealed facts or even religious experience. In good liberal fashion, Lindbeck concedes that "while there are of course nonreflective experiences, there are no uninterpreted or unschematized ones," and he concludes

---

28. James Smart, *The Interpretation of Scripture* (Philadelphia: Westminster, 1961), 250–51. See also Donald Gelpi, *The Turn to Experience in Contemporary Theology* (Mahwah, N.J.: Paulist Press, 1994), where the author traces this turn to the subject in a broad range of modern theologies: the New Testament christology of Edward Schillebeeckx, liberation theology, process theology, and transcendental Thomism.

by stating, "it is necessary to have the means of expressing an experience in order to have it, and the richer our expressive or linguistic system, the more subtle, varied, and differentiated can be our experience" (*The Nature of Doctrine,* 36–37). And if there is no pre-linguistic "real thing," just as there are no uninterpreted things-in-themselves, then another move toward realism was necessary.

Hans Frei is often credited with calling liberal theology to the task of finding its center. Thus, his call for a realistic reading of the Bible was both a return to a christological center and a reversal of a narrow, introverted approach to Scripture. Where modern hermeneutics went wrong was in trying to translate the Bible according to some external standard, some Enlightenment criteria such as reason or universal human experience. Rather than fitting biblical truth into a framework already constructed by philosophy, Frei countered by saying the biblical texts are the narratives that define the framework within which we rightly understand our experience and the rest of history. The locus of authority and revelation is the narrative, because it is in and through these stories and witnesses that we discover God's identity. This is a much broader foundation than the inerrant text where too much emphasis is placed upon the meaning of single words. And Frei revitalizes the tradition of the power of the word to create a new counter reality.[29]

For Frei the important thing is a realistic narrative, but not realistic in the sense that it meets scientific standards of truth or reality. The ontology of a realistic narrative arises from a plain or straightforward reading of Scripture as it supplies the interpretive framework within which believers live Christian lives and understand reality. The difficulty with this use of the word "realistic" is its introversion. Miroslav Volf provides a substantial criticism of Lindbeck's endorsement of Hans Frei's argument that theology

29. One of the important strengths of narrative truth is its ability to revitalize verbal truth without its usual association with verbal inspiration. See my discussion of narrative truth in Chapter 5. The innovative accent on "counter" is purely Brueggemann.

should be the redescription of reality from within the scriptural framework. Volf thinks Lindbeck is guilty of hermeneutical simplicity in the way he construes the influence between biblical and nonbiblical worlds as unidirectional. He points out what should be obvious for a nonfoundationalist like Lindbeck: "We can look at our culture through the lenses of religious texts only *as we look at these texts through the lenses of our culture....* Yet the whole program proceeds as if the religion that does the interpreting is itself not an interpreted religion" (*The Nature of Confession,* 51; italics his). Volf then offers his own metaphor: "Like some blimp fashioned out of a canvas of religious intersignifications, Lindbeck's cultural-linguistic system seems to float in mid air: we get into it and we see the whole reality anew from within it, and we behave differently because we are in it" (*The Nature of Confession,* 54).

Here there is a hint that at work is a romantic understanding of language where language expresses and represents an inner experience or cultural reality as if language is secondary to the real thing; the real thing being purely linguistic apart from its enactment. Lindbeck, however, decidedly rejects a romantic understanding of language (*The Nature of Doctrine,* 36–37). Therefore, one must conclude that it does not matter to him whether reality is realistic or factitious. It is difficult then to speak of a realistic narrative when the world arises solely from an intratextual reading of the Bible and when there are significant divisions on when to read a text literally or figuratively.

There are few theologians harder on the Frei-Lindbeck type of theologies than J. Wentzel van Huyssteen. Regarding insular theologies, he writes:

> However, when theological beliefs become a species of belief whose truth is discovered by means of criteria internal to the language game itself, this leads not only to a relativistic understanding of justification, truth, and knowledge, but to an epistemological relativism that would be fatal for the

> cognitive claims for theological statements—precisely in an age of postmodern scientific reasoning. (*Essays*, 89)

But why should the Christian interpreter care about a criterion that revolves around postmodern scientific reasoning? It matters to van Huyssteen because he sees rationality as an epistemological bridge between the disciplines of science and theology. Theologians have worked hard to establish their discipline as credible because it is rational. This was, and may still be, a necessary move. However, this approach ignores the ontological crisis that a realistic narrative must take into account when looking to a biblical narrative that has little relationship to realism as our culture experiences and understands it. Van Huyssteen is speaking from the methodological side of what constitutes valid criteria of justification; a separate and important issue that Frei and Lindbeck choose to ignore. A realistic narrative that derives all its meaning from the text is an incongruity. The meaning of the text may provide believers with a language and a set of concepts to make sense of the world, but the world should be recognized as kicking back.[30]

Theology of Soft Ontology and Soft Claims

I have spoken of the soft reality of a postmodern science. It is not surprising then to see in liberal theology a similar softening of its truth claims. As science moved away from Newtonian objectivity, liberal theology followed suit. Since the middle of the nineteenth century liberal theology has consistently reduced its ontological claims; that is, the content of the Christian faith has become smaller and smaller while authority was reassigned to epistemology. Rudolph Bultmann was a prime mover to bracket ontological claims and to reduce the content of theology to an existential decision. For example, John Macquarrie writes,

30. In the next section, I will develop the idea of reality as something that kicks back. For a similar assessment of Frei-Lindbeck's realism, see Jeffrey Hensley, "Are Postliberals Necessarily Antirealists?" in *The Nature of Confession*, 69–80.

> But although existential interpretation of a story does not in itself deny the factual content of the story, it certainly does put that content "in brackets," so to speak. The objective reference becomes bracketed in the sense that interest has shifted away from it to the existential significance. The question of fact is no longer being raised. We are not asking about what happened but about what the story says to us in our situation now.[31]

Karl Barth, on the other hand, took a very different course. Speaking as a scientist-theologian, Arthur Peacocke severely criticizes a theology that is self-insulating.

> [T]he neo-Barthian reliance on the pure Word of God available through the Scriptures has come to be seen by many as a non-viable route. For the further it takes us away from the presumed fallibilities of our natural minds to the supposed divine word in Scripture (or even in the Tradition), the closer it brings us to the question "How can we know that these Scriptures, or this Tradition, is transmitting to us the genuine Word of God?" (Peacocke, *The Sciences and Theology in the Twentieth Century,* xi)

This trajectory toward an ever more trivial ontology has now been reincarnated with contemporary twists. Sallie McFague's book *The Body of God* (1993) is an interesting example of how one theologian operates with a soft ontology. McFague appreciates that a complete rejection of ontology assigns us to relativism. She writes that when she uses metaphors of body, mother, lover, friend to describe God that these are meant to be ontological claims, "albeit shy ones." They are "shy" because there is no

---

31. John Macquarrie, *The Scope of Demythologizing* (London: SCM, 1960), 19. An interesting turn of events finds Walter Brueggemann doing the same thing. "Note well that in focusing on speech, we tend to bracket out all questions of historicity. We are not asking, 'What happened?' but 'What is said?' " See Brueggemann, *Theology of the Old Testament* (Minneapolis: Fortress, 1997), 118. This full and uncritical embrace of truth as speech (testimony) does concern me, as it does James Barr. See my further discussion in section 8, n. 50.

way to make authoritative claims for a correspondence between God's being and the world as it is or between our language and the way God is. "Models," she writes, "are to be judged not by whether they correspond with God's being but whether they are relatively adequate" (*The Body of God,* 152, 193). Behind this assertion is Niels Bohr's epistemological declaration that we cannot know the world as it is and therefore we cannot know God as God is. Since all models, whether theological or scientific, are not literal descriptions, McFague asks the reader to accept these various models of God's relationship to the world "as if" they were true.[32]

The softness of McFague's claims, then, are both epistemological and ontological. Her epistemology is also soft because there are no absolute standards of judgment. Therefore, "what we must ask of all models is their relative adequacy on the basis of some agreed-upon criteria" (*The Body of God,* 152). The criteria by which truth claims are to be judged as more adequate than alternative claims are those which have the broadest acceptance, such as, "the perspective of postmodern science, an interpretation of Christian faith, our own embodied experience, and the well-being of our planet and all its life-forms" (*The Body of God,* 152). I suspect the driving force behind her epistemology is the postmodern position expressed by the philosopher of science Mary Hesse.

> In our account of the interaction of cognitive systems of different kinds we need a quite different theory of truth which will be characterized by *consensus* and *coherence* rather than correspondence, by *holism* of meanings rather than atomism, by *metaphor* and symbol rather than literalism and

---

32. See *Religion and Intellectual Life* (spring 1988): 9–44 for a discussion between McFague, David Tracy, Gordon Kaufman, Rosemary Radford Ruether, James G. Hart, and Mary Jo Weaver, and especially Tracy's comments and McFague's response to the question whether an "as if" theology is adequate. Garrett Green, *Imaging God: Theology and the Religious Imagination* (San Francisco: Harper & Row, 1989) also discusses the postmodern condition when there are rival views of reality and each reality is construed as something.

> univocity, by intrinsic judgements of *value* as well as of fact. (quoted in *The Body of God,* 225, n. 36; italics in original).

The consequence of an epistemology driven by internal criteria, such as coherence and holism, is the proposition that there is to be no "reading off" evolutionary history. What follows is a complete rejection of a natural theology where "the scientific story gives evidence (even the tiniest bit) for belief in this or any other model of God and the world" (*The Body of God,* 65–66, 146, 160). McFague abandons discerning the conceptual "purpose or direction of divine activity" as it might be found in the universe and is merely content with an aesthetic sense of awe and wonderment.

If ontology counts for next to nothing, can there be a conversation with science? David Tracy and, more emphatically, Wolfhart Pannenberg, believe theology cannot dispense with the extra-textual reality of universal experiences and the universe, while Lindbeck argues no justification external or transcendent to one's religious interpretive framework is needed or possible. In Lindbeck's theological framing there is no inside-out or outside-in fit as such, only an appropriate cultural-linguistic expression for the living of one's faith. Lindbeck does have a "sort of" ontological truth; namely the correspondence or adequation between the lived Christian life (praxis) and the being and will of God (*The Nature of Doctrine,* 64–65). But it is weak.

> The latter locates religious meaning outside the text or semiotic system either in the objective realities to which it refers or in the experiences it symbolizes, whereas for the cultural-linguists the meaning is immanent. Meaning is constituted by the uses of a specific language rather than being distinguishable from it. Thus the proper way to determine what "God" signifies, for example, is by examining how the word operates within a religion and thereby shapes reality and experience rather than by first establishing its propositional or

> experiential meaning and reinterpreting or reformulating its use accordingly. (*The Nature of Doctrine,* 114)

There is no question that "to become a Christian involves learning the story of Israel and of Jesus well enough to interpret and experience oneself in its terms" (*The Nature of Doctrine,* 34). But just as existentialists bracket out the content of faith,[33] so Lindbeck has left us with no criteria to determine the truth of those stories. John Polkinghorne writes: "One is left uneasy. Why should anything be 'conceptually powerful' and 'practically useful' unless it bears some relation to the way things are?"[34] Even more damaging is the real possibility that something is conceptually powerful but false and even evil.

David Tracy makes the helpful observation that there are three publics that theologians address: the church, the academy, and society. Each public engenders a particular style or argumentation. "In house" speech among the baptized and interpreting the gospel so that it is relevant to a skeptical world are two very different functions, interests, and sensitivities. When you are preaching to the baptized, peculiar speech is appropriate (speech about repentance, sin, judgment, resurrection).[35]

The dilemma posed by Stout will not disappear by merely recognizing what public is being addressed. Because the Christian faith is primarily a message to be proclaimed, it cannot shed what is peculiar to it. Even when theology is primarily proclamation addressed to believers, it is, nevertheless, for the sake of the world. The question still arises: How does the scholar-theologian fit peculiar beliefs and peculiar ideas and a peculiar way of being-in-

---

33. I am reminded of the distinction made by Wilhelm Hermann (Bultmann's teacher) between the "experience of faith" and the "content of faith." Bultmann believed the experience of faith was sufficient to ensure the certainty of faith but what he did not properly understand, for it was never discussed in his lifetime, is the fact that there is not experience without some content that is "languaged" in the experience.

34. John Polkinghorne, *Reason and Reality* (Philadelphia: Trinity Press International, 1991), 14.

35. William H. Willimon, *Peculiar Speech: Preaching to the Baptized* (Grand Rapids, Mich.: Eerdmans, 1992); in the same vein, Stanley Hauerwas, *Dispatches from the Front: Theological Engagement with the Secular* (Durham, N.C.: Duke University Press, 1994).

the-world into speech and thought patterns that cannot possibly hold them? Christian scholars who sit at table with scientists are there as members of "truth seeking communities" (Polkinghorne's term); as such, they are not there to proclaim but to seek truth. Ellul and Brueggemann disdain literal or flat language because it is easily co-opted, homogenized, and flattened. Brueggemann prefers language that is always "elusively beyond the control of the rulers of this age."[36] The purpose of conversation is not just understanding by way of commensurability, but commensurability even where there is no agreement; that is, we understand each other, recognize the peculiarity of each other's language, and choose to differ.

Conclusion

Theology has struggled valiantly to define what is distinctive about its methodology and has done so, oddly enough, with only scant attention being paid to inspiration and revelation. This is not to say there has been little or no discussion about inspiration and revelation, only that it has happened within a very narrow context, such as inerrancy, and seldom as part of the larger philosophical discussion of epistemology-ontology. Theology's attempt to emulate empiricism has left no one content. As long as a Newtonian style of science held sway, theology emulated a model of correspondence presuming this would provide a firm foundation. It was not to be! The physical sciences and philosophy were going in different directions after centuries of mutual support. A new twist had been added to the eighteenth-century split between truths of reason (the domain of theology and philosophy) and truths of facts (the domain of science). The conclusion that soon followed was: If one is false, then the other is true; if one is not meaningful, then the other is; if one is real because it is true,

36. See Brueggemann, *Cadences of Home* (Louisville: Westminster John Knox, 1997), 58; and Ellul's *The Humiliation of the Word* (Grand Rapids, Mich.: Eerdmans, 1985), 22–26. I am also thinking of Friedrich Nietzsche's quip that truth "is an army of metaphors."

then the other is true because it is real. Science would try to proceed without metaphysics by reducing its truth claims to verifiable statements of facts. Theology, in spite of a sustained effort, could not forge an epistemology of verifiable facts. Nor could it fashion one of clear and distinct ideas.

Theology still had the option of asserting a supernatural ontology, but that battle had been lost in the minds of many by the time of Darwin's death. Conservative theology, nevertheless, remained committed to an understanding of language as fitting together the facts of revelation and history with an inspired word.[37] But how do we decide which reading is the correct one? Liberal theology reluctantly turned its back on the hard ontology of Newtonian science and found an alternative in the ontology of experience and praxis. Could this alternative between chasing science or chasing philosophy be the reason why theology itself became a house divided: divided between conservatives who want to believe, as good rationalists using the methodology of science, in the legitimacy of a literal (tight) fit between word and God, and liberals who accept the tenants of a postmodern science that we can only know something tentatively and approximately (loose fit)? Is this the background behind those who defend revelation as the naming of God and those who will settle for, because it is the best we can have, metaphorical forms of language about God?[38]

We have not moved very far from the Barth-Bultmann debates about the proper starting point of theology: the Bible or the world. This is a false and artificial dichotomy because we experience and know God through a continuous play between world and language, experience and Scripture. The hermeneutical circle is inevitably in-out and out-in. The critical issue is not the starting

---

37. Rodney Clapp, "How Firm a Foundation: Can Evangelicals Be Nonfoundationalists?" in *The Nature of Confession,* 81–92. His conclusion is that many evangelicals still hold to the mood and rhetoric of foundationalism and that in the end it "comes down to who is reading the Bible correctly" (87).

38. I have in mind the inclusive language debate. See J. A. DiNoia, who carefully unpacks these two ways of how we might speak of God: "Knowing and Naming the Triune God: The Grammar of Trinitarian Confession," ed. Alvin F. Kimel Jr., *Speaking the Christian God* (Grand Rapids, Mich.: Eerdmans, 1992), 162–87.

point but whether the hermeneutical circle is completed. Should some kind of priority be given to Scripture? Yes, but not as a description of reality based upon a "plain" reading of the texts.[39] Lindbeck and Frei ask too much of the biblical texts. They are not capable of providing a narrative framework without considerable interpretation and certainly not a schema that can absorb a modern scientific understanding of the universe. We cannot do what Frei wants us to do; namely, to start with "the biblical world" and let these narratives define what is real, because it will always come down to which reading of the Bible is the correct one. Scripture alone cannot be the foundation because it contains texts that, if accepted as self-evident, run counter to the larger, overarching proclamations of a loving and just God. Feminist readers of Scripture Elisabeth Schüssler Fiorenza and Phyllis Trible are appalled at the prospect of allowing a literal or plain sense reading of certain texts to "absorb all other worlds."[40] An intratextual reading would leave no place for the experience of women struggling for liberation from patriarchal oppression, and this constitutes a serious defect for any epistemological approach that limits the hermeneutic circle. Clearly, the hermeneutical circle—the fit between world and Word—is difficult to maintain. It needs to be seen if the paradigm of narrative truth will prove to be a sufficient remedy.

---

39. For an explanation of a "plain sense reading" see Frei, *The Eclipse of the Biblical Narrative* (New Haven, Conn.: Yale University Press, 1974) and Kathryn E. Tanner, "Theology and the Plain Sense," in *Scriptural Authority and Narrative Interpretation,* ed. Garrett Green (Philadelphia: Fortress, 1988), 63.

40. See Elisabeth Schüssler Fiorenza's *Bread Not Stone: The Challenge of Feminist Biblical Interpretation* (Boston: Beacon, 1984) and Phyllis Trible's *Texts of Terror: Literary-Feminist Readings of Biblical Narratives* (Philadelphia: Fortress, 1984). Both raise methodological questions when they encounter the Bible "as a thoroughly patriarchal book written in androcentric language." Both find it necessary, therefore, to reject a plain-sense reading while moving toward a broader view of inspiration and a narrative reading of the texts.

– SECTION 8 –
THE REAL THAT INCLUDES QUESTIONS OF TRUTH

The conversation between science and theology should be about what is real and our place in it. What merits the attention of both scientists and theologians is constructing the narrative truth about the kind of universe we live in, the kind of creatures we are, and our purpose in the universe. On the part of theology, such a conversation will require a turn to the real; on the part of science, a larger vision of what constitutes truth.

There are at least two kinds of truths. The first kind is the fitting together of cognitive thoughts with the external real of the universe. The second kind is verbal truth, which does not necessarily depict or explain the real of the universe but tries to answer perennial questions that arise from our human capacity for self-transcendence. One could think of these two kinds of truths as language-to-world and language-to-language truths. A shorthand expression is to speak of them as empirical truths and word-truths. Narrative truth, then, becomes the catalyst to bring together truths that point to the real as actually important and truths that are language itself trying to state what is meaningful.

Empirical truth claims and language truth claims are very different and give rise to quite different ontological-epistemological fits. Furthermore, truths of physical reality and truths of language have a veracity of their own. We have caused ourselves no end of difficulty by giving priority to one kind of truth and by denying that either kind of truth can have its own moments of transcendence, creativity, and discovery apart from the other. Although it is possible, and at times helpful, to ask the question and seek the answer solely within the framework of language or physical reality, it is a serious detriment for the conversation between theology and science.

In *The Humiliation of the Word* (1985) Jacques Ellul provides a perspective that is crucial to this discussion. Ellul may be the most consistent philosopher-theologian when it comes to recognizing the effect the dominance of science has had on society.[41] Ellul argues that science has succeeded in convincing most of us that the only possible truth consists in knowing reality, and that the only evidence of truth is success relative to reality, for example, how well one can manipulate "things." To be more specific, truth has become so identified with reality that we have forgotten "the word belongs to the order of the *question of truth*. An individual can *ask* the question of truth and attempt to *answer* it *only through language*" (*Humiliation*, 29, italics in original). Ellul believes that meaning and understanding arise from how we use words and that it is the use of language that permits us to go beyond the reality of the external.

Ellul's analysis raises two questions: What happened to the truth of the spoken universe? Is there no place for the unique value of language to be an equal partner in judging what is true and what really matters? Ellul makes no apology for the fact that the truth of language is multidimensional, rich with connotations, textured with many levels of meanings, and is filled with a rich complexity of things left unexpressed (*Humiliation*, 16–26). The power of language, then, is its freedom not to be bounded by the physical universe. Ellul asserts that in contrast to reality, language never belongs to the order of evident things but is in continuous movement between hiding and revealing. "The blessed uncertainty of language is the source of all its richness," for it bristles with connotations and evokes a myriad of images. Therefore, language is unique by enabling us to stand apart and question. Language is the necessary assertion between the verifiable universe and the self; it prevents us from collapsing what is real into what is true, or what is true into what is real, since

41. Ellul's critique of modern technology stretches a long way: *The Technological Society* (New York: Knopf, 1964), *Propaganda* (New York: Knopf, 1965), *The Political Illusion* (New York: Knopf, 1967), *The Technological Bluff* (Grand Rapids: Eerdmans, 1990) and many other books and articles.

the human self is in that ambivalent state of being part of the material universe and yet separate from it.

Moving within this framework, Ellul continues to contrast the distinction between reality and truth. He explains why, in the narrow sense, scientific descriptions of reality can be accurate or inaccurate but not false or true. Reality can be reified, computerized, and endowed with ultimate value. The true, on the other hand, is known through subtle nuances and fragile transmissions. While the physical reality of existence is indispensable and useful, providing the furniture for existence, it provides us with no direction to follow. "It leaves us with fragments of this reality, to flounder as best we can without a compass or a sextant, in the midst of the continually shifting waves of this world" (*Humiliation*, 229). "Reality can be obvious, but truth never is" (34).

Ellul's assessment of the modern world includes the observation that we have tried to restore truth through sight. "Our era," he writes, "is further characterized by an absolute identification of reality with truth" (31). In a culture where all truth is contained within reality and that reality is confirmed by what we can see and manipulate, the image becomes the most real. What people put their confidence in today, Ellul declares, is "not on the sight of a reality that surrounds them, but rather on the multiplication of artificial visual images that constantly attract their attention" (192). We have shifted our trust onto images of reality, such as the video images of the screen and tube. We bow down before the representation of reality; the image becomes unquestionable until the image becomes more real than reality itself. In this very process, we confuse truth for reality, believing what is true excludes what is discrete and hidden. A visualized reality is a controlled reality.

Paradoxically, every writer/speaker must put some trust in word-truth in order to argue for truth. To the extent that science moves to reduce empirical truth to the best argument available and uses words to decide which theory or hypothesis is the better one, then the truth of words is given ever more weight. As science becomes fonder of concepts and paradigms that are philosophically charged (mass, causality, complementarity, field, chaos, etc.), its words are less connected to the ontological real and more to the thread of argument.

Succinctly stated, there are at least four reasons why the real must be put back onto everyone's academic agenda and why the conversation between theology and science must give particular attention to the real that includes questions of truth.

First, the real informs epistemology and thereby changes it. The interplay between epistemology and ontology requires that both be attended to simultaneously. With the nearly total emphasis on the importance of methodology in our era and the subsequent eclipse of ontology, the interplay has become one-directional. What we hear repeatedly is that what we discover is the consequence of the methodology we choose; what we find is dependent upon what we are looking for. The other direction of influence, however, is from the side of reality when it provokes a new methodological approach. Like a tennis match, as the epistemic ball is served it is not always returned as expected or predicted; so, if we are attentive, we choose to alter the way we serve the next time.

Time and again scientists find that reality does not kick back in just any way but in congruence with the language of mathematics; that is, there is an intelligibility—as Einstein insisted—built into the universe. This built-in intelligibility explains why we can talk about roughly the same world and why there is a certain historical continuity, or what Arthur Peacocke refers to as "the involvement of scientists in a continuous linguistic and experimenting community" (*Intimations of Reality*, 33). There is the increasing ability to predict and control entities.

Freeman Dyson observes the time has come to redress Thomas Kuhn's interpretation of scientific history. Published in 1962, Kuhn's *The Structure of Scientific Revolutions* was an important and needed reassessment of a hard science claiming too much. As a typical postmodernist, Thomas Kuhn was so preoccupied with the state of methodology that he scarcely considered the possibility that science was compelled to make softer claims because it had encountered a different and softer reality than Newton had. Kuhn would have us ask if this change in perception is anything more than a paradigm shift and how we know that this time we have discovered the real universe. Dyson counters that "some scientific revolutions arise from the invention of new tools for observing nature" and "others arise from the discovery of new concepts for understanding nature."[42] Dyson points to Peter Galison's recently published book, *Image and Logic* (1997). The contrast between the two is that of science driven by new concepts (paradigms) and science as a process of discovery by way of new tools. Galison's book is filled with hundreds of pictures of scientific apparatus, while Kuhn's book has none. Dyson also observes that most of the recent scientific revolutions have been tool driven, such as the double-helix revolution in biology and the big-bang revolution in astronomy. This is not to demean concept-driven revolutions; however, it does reflect a position of realism that accommodates both discovery of the mind as well as of the eye.

Second, the real serves as a check against speculation. When Francis Bacon leveled the charge of speculation against theology, he was primarily referring to the way theologians were making truth claims with no observational, demonstrable basis. Although religious experience had an aura of directness and immediacy, it seemed quickly to take flight into heavenly matters. Theology chose not to compete with modern science on a language-to-world level, so was left to erect the ramparts defending the veracity of verbal truths. It was unclear, though, how theology was to defend itself against the charge of speculation, but in the

---

42. Freeman J. Dyson, *The Sun, the Genome, and the Internet: Tools of Scientific Revolutions* (New York: Oxford University Press, 1999), 13. Dyson comments, "Unfortunately, Kuhn's version of history was dominant for thirty years before Galison's version [*Image and Logic*] appeared to restore the balance" (14).

heat of the historical moment speculation became synonymous with having no external reference. From a postmodern perspective, we understand the simplicity of Bacon's argument and are willing to admit a variety of truth claims arising from a variety of observational claims.

Third, John C. Polkinghorne states that science encounters a universe that transcends its own power to explain.[43] The observation could also be stated in this way: the universe contains data that possess dimensions with theological relevance. After several centuries of trying to eradicate metaphysics, there is a new willingness to raise and wrestle with meta-questions. The anthropic principle is just one example. How does one explain a universe that evolved in such a way that chance and necessity did not cancel each other, producing nothing or something that is thoroughly sterile, but instead gave rise to an astonishingly finely tuned universe capable of supporting systems of such fruitfulness and complexity that we are here today reflecting upon what has happened? Such questions go beyond a verifiable universe. Is it unreasonable to assert a universe that does not come into existence solely because it is described in words and yet is another universe because it is described in words?

Finally, it is not only that reality provokes meta-questions and questions we cannot ask of ourselves, but that the real has the possibility of telling us something about God and God has the potential of revealing something about the universe. Wolfhart Pannenberg has been very consistent in requiring an unequivocal connection between God and the universe.

> If the God of the Bible is the creator of the universe, then it is not possible to understand fully or even appropriately the processes of nature without any reference to that God. If, on the contrary, nature can be appropriately understood

43. While this idea is found in several of his books, the most pointed formulation is in "Creation and the Structure of the Physical World," *Theology Today* 44 (April 1987): 53–68.

> without reference to the God of the Bible, then that God cannot be the creator of the universe, and consequently, he cannot be truly God and be trusted as a course of moral teaching either. (*Toward a Theology of Nature: Essays on Science and Faith*, 16)[44]

Several interesting questions arise. What ontological events would be compatible with a created universe? With a universe created by a personal God? Is life in all its fullness and depth realizable apart from God as creator, redeemer, and sustainer? Is the hypothesis of God ever a necessary component of the better explanation? Can we possibly understand our place in the universe, including why we exist as self-reflecting, worshiping creatures, without the wisdom of tradition, holy texts, religious experience? When we restrict our conversation to finding epistemological parallels, there is the inevitable danger that we will ask and hear only ourselves speaking.

Definition of Realism

The reader is entitled to my understanding of realism, since it is my contention that it must be a cornerstone in any conversation between theology and science. I begin by accepting both "an objectivity which is already constituted universally and prior to experience" (Habermas) and our varied, colorful, individual experiences and interpretations of that universe. The "objectivity" of the universe does not mean we have access to things-in-themselves, only that there exists a reality independent of the human mind. This implies that we cannot define truth strictly in epistemic terms. Eberhard Herrmann sets forth a fundamental dictum that runs counter to much of contemporary philosophy:

---

44. It should be remembered that Pannenberg was one of the first major theologians to seek a middle way between Barth and Bultmann; that is, between a top-down thinker (Barth) and a bottom-up thinker (Bultmann). This initial thrust can be seen in his early essays on history and revelation. See W. Pannenberg, ed., *Revelation as History* (New York: Macmillan Co., 1969).

"The fundamental realist intuition consists of the thesis that the world is objective or real in the sense of being independent of how we think of it. This implies that we cannot define truth in epistemic terms."[45] The other side of the coin is stated in the strongest language possible by theoretical physicist David Deutsch:

> We realists take the view that reality is out there: objective, physical and independent of what we believe about it. But we never experience that reality directly. Every last scrap of our external experience is of virtual reality. And every last scrap of our knowledge—including our knowledge of the non-physical worlds of logic, mathematics and philosophy, and of imagination, fiction, art and fantasy—is encoded in the form of programs for the rendering of those worlds on our brain's own virtual-reality generator. (*The Fabric of Reality,* 121)

A realist approach allows us to believe that we can have methodologies that come ever closer to approximating the world as it is, while still affirming that the ontology of the universe is such that it will always be too complex and ever evolving to be described fully in either words or symbols.

David Deutsch, whose papers on quantum computation laid the foundations for that field, provides a minimal definition. He writes: "We must adopt a methodological rule that if something behaves as if it existed, by kicking back, then one regards that as evidence that it does exist. Shadow photons kick back by interfering with the photons that we see, and therefore shadow photons exist" (*The Fabric of Reality,* 88). This position is compatible with functionalism, which emphasizes the qualities of experimental manipulation and historical continuity. Ian Hacking observes that in principle even entities that cannot be observed are regularly manipulated to produce new phenomena. "By the time that

45. Herrmann, "A Pragmatic Approach to Religion and Science," in Niels Henrik Gregersen and J. Wentzel van Huyssteen, eds., *Rethinking Theology and Science* (Grand Rapids, Mich.: Eerdmans, 1998), 130.

we can use the electron to manipulate other parts of nature in a systematic way, the electron has ceased to be something hypothetical, something inferred."[46] Deutsch's minimalist position is not willing to speak about truth, as in language-to-world truths, or about a fit between language and entities and their essential properties. Deutsch, for instance, argues that observational evidence can be used to prefer one theory to another, but it is not evidence "read off" reality as to its true nature. Observe how he qualifies his argument: "As I have said, it is impossible literally to 'read' any shred of a theory in nature; that is the inductivist mistake" (94). But why must it be a "literal" reading and is there no shred of evidence that points to the otherness of reality?

John Polkinghorne, with advanced degrees and experience in both science and theology, has been a consistent advocate for a moderating view. Writing about the acceptance of the standard model of the quark theory of matter, he speaks of the "immense struggle to find a theory that is economic and uncontrived and adequate to a wide swathe of experimental investigation" and that when a coherent picture emerged, "it had all the feel of discovery and none of the feeling of pleasing construction" (*Beyond Science,* 9). The result was a communal exclamation: "So *that's* what nature is like—who'd have thought it beforehand!" (*Beyond Science,* 9).

From a very different perspective, C. S. Lewis discovers what is real as it stands over and against memory. Lewis married late in life and his wife Joy awakened feelings that had been dormant. He knew that she had cancer while courting her and probably felt, as all lovers do, that love would conquer all. A slim book entitled *A Grief Observed* is his very personal reflection of one who had "lost his world" and faced a future without his Joy. Lewis questions the value (and pain) of remembrance:

---

46. Ian Hacking, *Representing and Intervening,* 262. The distinction that Hacking presses is between being a realist about entities and about theories. He believes the former can be defended, while the latter cannot.

> Slowly, quietly, like snow-flakes—like the small flakes that come when it is going to snow all night—little flakes of me, my impressions, my selections, are settling down on the image of her. The real shape will be quite hidden in the end. Ten minutes—ten seconds—of the real H. [Lewis's affectionate name for his wife] would correct all this. And yet, even if those ten seconds were allowed me, one second later the little flakes would begin to fall again. The rough, sharp, cleansing tang of her otherness is gone. (*A Grief*, 21–22)

During this time of grief Lewis came to understand that the most precious gift that marriage gave him was the constant impact of something very close and intimate yet all the time unmistakably other, resistant—in a word, real. And so he cries out, "Oh my dear, my dear, come back for one moment and drive that miserable phantom away" (*Grief*, 20). The phantom was his memory of her. His memories are the reason for his grief. These memories, all that he has left, he knows will sift out like sand falling through a sieve until they too are gone, but there would be no memories if there had not first been the otherness of Joy.[47]

A Robust Ontology

I wish to go beyond this "bare bones" position by arguing for a universe that has definite ontological properties and therefore presents something implicitly that can be made explicit by words and mathematics. A robust ontology is one that not only kicks back, but one that has a history which verges on narrative. Once again, we are confronted with the question: To what degree does

---

47. C. S. Lewis, *A Grief Observed* (New York: HarperSanFrancisco, 1989). As vivid as the experience of the recreated Jesus can be, and some experience Jesus as more vivid than life itself, it is at best momentary, soon (ten seconds later) the snowflakes of our memories, our selections, and our impressions begin to fall. This explains the different accounts of the risen Christ as it is recorded in the four Gospels. The real, the otherness of Jesus as they knew him in the flesh, is thereby affirmed, because otherwise there would have been no longing for one more moment with the Jesus they once knew. This is vividly told in John's account of the encounter of Jesus "behind locked doors" (John 20:19–29).

the independent world impress itself on conceptualization? The Kantian notion of a conceptual fit between a priori (universal and given) concepts and the world was naïve. Postmodernists like Rorty are right that any fitting resembling the Cartesian model is static and therefore artificial, but the ontological baby that was thrown out cries out to be heard.

A balanced view of realism must include the ontological givenness of the universe, the subjective conceptualization of the mind, and cultural conditioning and practices. In order to achieve this balance, the inflated role of the self must be reigned in; otherwise, "otherness is swallowed up into cultural and hermeneutic practices."[48] Words like "fit" and "correspondence" may be inappropriate, nevertheless they point to a mutual interplay between mind and world. Farrell emphasizes an understanding of the world as constraining and regulating our conceptual schemes of it. "The important idea here," Farrell writes, "is that reality is not the child of our abilities for organizing, meaning, and evidence gathering, but is an independent measure of how good those abilities get to be" (*Subjectivity*, 149). At one point he speaks of reality as "educating" our sensibilities toward it (*Subjectivity,* 169). Is it so far-fetched to believe once again that nature has something to teach us? Both words, "impress" and "educate," are active, and this is the tense we must regain because it catches the dynamic and holistic way the world and our ways of thinking about it evolve in tandem.

In a robust ontology, methodology does not determine ontology—nor does ontology determine methodology. Language cannot be easily fitted to the world because the world is not sentence-like. On the other hand, the world is not devoid of its own independent meaning, ontological hooks, and decisive events. As a viable ontology is reconstructed, we must assume the real is materially important but not particularly meaningful

48. Frank B. Farrell, *Subjectivity, Realism and Postmodernism* (Cambridge: Cambridge University Press, 1996), 250.

without different methodological techniques utilizing language in distinctive ways.

In deciding between a position of strong realism, weak realism, and a balanced position of critical realism, the pivotal issue is how undifferentiated the world is prior to our conceptualization of it. A strong position would like to take the best of Aristotle and Kant and argue that the world has an ontological structure and human minds have conceptual structures and the two can be fitted together in an objective way to tell us the way the world is. A weak or soft realism argues that there is no determinate sense to the way words become attached to things we encounter outside of ourselves. The realism I have in mind has "ontological hooks," not in the sense of a DNA template, but places that call forth metaphysical debate, conceptual consternation, and religious inference.[49] I would again caution against the "all or nothing" package. Frank Farrell is worth quoting another time.

> We do not want the world to be just a potential and indeterminate measure of truth that comes into play once our activity of individuation has marked off lines in it. A realism robust enough to deserve the name should specify more of a back-and-forth process through which our way of articulating must be to an important extent self-articulating so that even if it offers more than one option as to how we might raise some of its articulations to metaphysical prominence, depending upon our interests, still it must impose a significant discipline on our work of individuation. (*Subjectivity, Realism,* 163)

Theological Realism

A theological realism embraces two kinds of truths. On the one hand there is Ellul's spoken universe; on the other, the empiri-

49. The phrase "ontological hooks" is Farrell's. In Daniel C. Dennett's understanding of evolution, such ontological hooks become "skyhooks"—a disparaging term referring to thrusts of special creation and ad hoc hypotheses used to explain what I am calling "decisive events." See Dennett, *Darwin's Dangerous Idea,* 76, 83.

cal truths of science. Sometimes we consult the contours of the world to test the truth of our semantic utterances and at other times we do not. A theological realism acknowledges the power of words to construct a counter-reality, for we are always in danger of attaching ultimate truth to the wrong thing.

The theological realism advocated here is not modeled after a scientific realism where consensus and progressive approximation (prediction and manipulation of entities) is paramount; it is not motivated by increasing commensurability as it is with van Huyssteen (mutual reasoning strategies) or Nancey Murphy (methodological convergence); it does not pursue a new form of mutual corroboration; and it is not a way to construct an integrated worldview. On the other hand, a robust theological realism returns theology to what is real, does not allow theologians to perpetuate a situation where epistemological concerns dominate, protects theology as a word-truth enterprise from speculation, provides a counterweight to those naturalistic explanations that presume to be entirely adequate (á la Richard Dawkins, *The Blind Watchmaker*), and engages theology with the perennial questions of life in the universe.

Theological realism is about stories and narratives. If there is a story to be told, it is because we believe that some things that happen do not just happen like leaves being blown off a tree by the wind. There is order and purpose, deep down and behind perfunctory sight, that leads us to think life somehow adds up, that life, and even the universe, is like a story. Can we clearly or neatly separate what is chance and providence, what is objective and what is subjective? No. Theological realism, however, is willing to try, because the narrative character of the universe invites us to try. Theologians would have us consider the proposition that God created the universe so that we might discover what is true, beautiful, courageous, and faithful.

Theological realism is ultimately about word-truth. Old Testament scholar Walter Brueggemann has been developing a similar paradigm over the years, fully developed in his *Theology of the*

Old Testament (1997). In his assessment of the contemporary situation, he traces the collapse of a historical reading of Scripture where history was considered as moving in a developmental line and where an objective investigation could recover history "as it happened." The "facts" and the "mighty acts of God" were "innocently available, so also the 'meanings' embodied and enacted in those facts" (47). We see the shadow of positivism and empiricism of trying to fit theological articulations into historical, objective containers. The world behind the text is not available to us; nor was it the intention of the authors to make it primary. Brueggemann understands the Old Testament in terms of theological rhetoric *"capable of construing, generating, and evoking alternative reality"* (59, italics in original). Speech, in particular testimony and counter-testimony, constitutes the new reality, an alternative reality to the hegemonic powers Israel faced in exile (65–68). Brueggemann acknowledges that his notion of "speech as constitutive of reality" is problematic in the sense that God can quickly become "a mere rhetorical construct" (65, n. 11). This is certainly a problem when so much emphasis is being placed on the role of rhetoric and intertextuality as constitutive of a theological reality.[50]

Conclusion

This book was never intended to be a full-blown apology for ontology, but I certainly want a place in my theology for a vigorous notion of an ontological givenness about reality that can be read in a limited way. In everything that I have asserted, my ontology is hesitant and my epistemology even more so; as a whole,

50. James Barr fears that Brueggemann, in his deference to postmodernists such as Foucault and Derrida, has evaded the ontologically Real and succumbed to a "remarkably non-historical approach." He rightfully notes that in his weighty *Theology of the Old Testament* Brueggemann gives scant attention to the narrative texts, people, historic rhetoric and mostly ignores historical-critical questions. This, it seems, is theological rhetoric in excess. See Barr, *The Concept of Biblical Theology: An Old Testament Perspective* (Minneapolis: Fortress, 1999), 545–62.

however, it is a counterweight to the hard ontology of some and the nonexistent ontology of others. By no means do I read history, bio-history or any kind of history, as a straight line, but I do understand history as having a storyline; that is, a drama driven by decisive events. When these events are brought to expression by a human mind with the capacity for self-transcendence, holistic patterns begin to emerge. This narrative storyline is not a metahistory nor is it made up as we go along. The givenness of the universe is ambiguous and oblique; to make it meaningful, the intelligence of the mind and the faith of the heart are required. No apology is needed for what the human species reconstructs, as if it were in some way bogus or illegitimate.

It is futile to bind up the fragments of foundational pillars that have either been thrown down or fallen down. The comforting sense of place in a divinely goal-imbued cosmos—along with the remnants of a Newtonian worldview—are not recoverable. To pursue this course would tempt us to hang onto fundamentals, forgetting Wolterstorff's definition of foundationalism as "a body of propositions which can be known by the natural light of reason" (*Within the Bounds of Religion,* 30). A revival of ontology must contend with the fault line that makes us choose between word-truth and fact truth, between the givenness of the universe and the creativity of human thought.

We have rightly distanced ourselves from the Enlightenment belief in an ideal fit and understand it as naïve, but are we justified in denigrating both language and object to the point that we dismiss either or both?[51] The denigration has taken place on at least two fronts: by squeezing the givenness out of reality and by treating the truth of language as if it were no truth at all. Just as the eclipse of ontology was spawned by the separation of the

---

51. Richard Rorty, typical of the postmodern attitude toward a robust ontology, remarks: "facts, like telescopes and wigs for gentlemen, were a seventeenth-century invention." See Rorty, *Philosophy and the Mirror of Nature* (Princeton: Princeton University Press, 1979), 357. His statement is true enough, but do we therefore throw out the notion of fact?

word-truth of theology and the empirical truths of science, so its recovery will depend upon reconnecting the two. We must defeat the present fissures that dominate—the truth is either constructed from the subjective side as verbal or it is disclosed (language of revelation) or discovered (language of science).

Chapter 4

The Conversation That Makes the Most Sense

The misguided search for intrinsic meaning within nature—the ultimate (and also oldest) violation of NOMA [non-overlapping magisteria]—has taken two principal forms in Western tradition. —STEPHEN JAY GOULD, *Rock of Ages,* 178

Taken by itself the theory of coherence proves insufficient to account for the most strongly held claim of religious truth—namely, that it originates outside of the system.

—LOUIS DUPRÉ, *Religious Mystery,* 33

– SECTION 9 –
THREE MODELS CONSIDERED AND FOUND INADEQUATE

Introduction

While there are many variations, there are only three basic approaches to the conversation between science and theology. The two disciplines represent parallel lines, converging lines, or lines that intersect at various points or levels. Ian Barbour's book *Religion in an Age of Science* (1990) described the alternatives as conflict, independence, dialogue, and integration. John F. Haught's *Science and Religion* (1995) uses the fourfold typology of conflict, contrast, contact, and confirmation, preferring the latter two. Haught does not find sufficient clarity in Barbour's dialogue and integration models, so collapses them into what he calls contact and then adds a fourth type, "confirmation." Since Barbour's seminal work, a new vocabulary has entered the discussion: consonance, consilience, complementary perspectives, and graceful duet. The basic alternatives, though, have not really changed.

Parallel lines assume that science and theology have dissimilar and incompatible epistemologies. Little is said to the effect that they may also have quite different ontologies, constituting a different fit. It follows that each would operate with distinctive and incommensurate ways of coming-to-know and would use language in very different ways. Furthermore, they would have distinctive governing interests: interpretation and explanation of religious experience leading to a religious way of life instead of interpretation and explanation of the nature of the physical universe leading to universal laws. They each work with data that would seem domain specific: religious feelings instead of experimental data; historical events instead of mathematical equations; "solvable problems" instead of "unsolvable mystery" (Haught). Therefore, it would be better for both if they respected the integrity and uniqueness of their own particular interests and methodologies.

Converging lines imply integration of some kind. Since Teilhard de Chardin's attempt to integrate science and theology, almost everyone is suspicious of any form of harmonization. We have become more sophisticated with words like consonance and coherence.[1] Converging lines imply a degree of corroboration. When the conversation aims at convergence, theology and science are considered cousins under the skin because they both look to provide a "coherent and deeply intellectual satisfying understanding of the total way things are" (Polkinghorne, *Reason and Reality*, 51). While integration may be too strong a word, and hypothetical consonance too weak, this view of the conversation is committed to a larger framework utilizing insights from all disciplines, especially from philosophy.

How theological and scientific lines intersect suggests a third possibility. The distinction is that in place of coexistence or convergence, theologians and scientists probe points of intersection. There is what Polkinghorne calls "cross trafficking," what Gilkey names "the nexus of science and religion," and what Barbour refers to as "indirect interactions." "Contact," John Haught's term, "insists on preserving differences, but, it also cherishes relationship" (*Science and Religion: From Conflict to Conversation*, 18). Wolfhart Pannenberg comes closest to the direction that looks the most promising. The conversation proceeds on the assumption that the deep intelligibility of the natural world increases when the universe is understood in its relationship with

1. For an explanation of consonance and its application, see Ted Peters, ed., *Science and Theology: The New Consonance* (Boulder, Colo.: Westview, 1998), 1–2, 11–39. For the development of coherence as a model, see Niels Henrik Gregersen, "A Contextual Coherence Theory for the Science-Theology Dialogue," in *Rethinking Theology and Science: Six Models for the Current Dialogue*, 181–231. John Polkinghorne also discusses consonance over against assimilation as two alternative approaches as it is developed in the writings of Ian Barbour, Arthur Peacocke, and himself. See Polkinghorne, *Scientists as Theologians* (London: SPCK, 1996). For an earlier discussion of consonance, see E. McMullin in A. R. Peacocke, ed., *The Sciences and Theology in the Twentieth Century* (Notre Dame, Ind.: University of Notre Dame Press, 1981), 17–57, especially his conclusion.

God. And the converse is true: our understanding of God increases when we know more about the nature of the universe. "The reality of God," he writes, "is a factor in defining what nature is, and to ignore this fact leaves something less than a fully adequate explanation of things" (*Toward a Theology of Nature,* 16, 48). The driving questions become: What will the real teach us about God? And what will the universe teach us about the nature of God? Intersecting lines imply a critical distance so that collaboration does not mean mutual verification or seeing with the same eye, so much as it means struggling with the same perennial questions and finding sufficient consensus to encourage action regarding the future of our planet.

I will examine three positions based on how they approach the epistemological-ontological fit that I have deemed a critical tool in order to understand our situation. Inherent in this discussion is how each position treats language and its relationship to reality. Each is a model with a certain goal in mind: separate but valid domains, converging methodologies, and intersecting interests. As might be expected, each approach has its particular strengths that commend themselves to us.

Stephen Gould and Willem B. Drees are advocates for a functional pragmatism arising from separate but equal domains. Theology and science complement each other but function best within their own separate domains. Drees writes, for instance: "I do not mean that they together result in a complete view, but I suggest that we see them as independent contributions that can be brought together in a larger worldview."[2] Gould draws a firm line when he writes:

> This book rests on a basic, uncomplicated premise... NOMA [non-overlapping magisteria] is a simple, humane, rational, and altogether conventional argument for mutual respect, based on non-overlapping subject matter, between

---

2. Willem B. Drees, "Postmodernism and the Dialogue between Religion and Science," *Zygon* 32, no. 4 (December 1997): 539.

> two components of wisdom in a full human life: our drive to understand the factual character of nature (the magisterium of science), and our need to define meaning in our lives and a moral basis for our actions (the magisterium of religion).[3]

The most consistent proponent of methodological convergence, or consonance, is Nancey Murphy. The most emphatic expression of convergence is her argument that theology, methodologically speaking, be indistinguishable from the sciences (*Theology in the Age of Scientific Reasoning,* 198). The strong pole of this alternative is convergence and the weak pole is consonance. The essential argument is for theology and science to find places where they corroborate each other concerning shared areas of correspondence. Simply stated, reality that is discerned both scientifically and theologically yields greater illumination and understanding.

"Points of intersection" is a broad description of the third alternative. One thinks of Wolfhart Pannenberg's *Toward a Theology of Nature* as the seminal work probing areas of possible intersection. This third alternative has certain affinities with the second in that it is interested in collaboration but less concerned about corroboration. The culminating work of J. Wentzel van Huyssteen, *The Shaping of Rationality*, is perhaps the most sustained argument for rationality as that indispensable place of contact.[4]

---

3. Stephen Jay Gould, *Rock of Ages: Science and Religion in the Fullness of Life* (New York: Ballantine, 1999), 175. Gould is not the only well-known scientist who feels separate is better. See Richard Dawkins, *Unweaving the Rainbow* (Boston: Houghton Mifflin, 1998). Edward O. Wilson is of an opposite mind, as is obvious by the title of his book, *Consilience: The Unity of Knowledge.* Wendell Berry is particularly critical of *Consilience* because he sees in Wilson the move to crown science as the new sovereign. See Berry, *Life Is a Miracle* (Washington, D.C.: Counterpoint, 2000), chap. 3.

4. J. Wentzel van Huyssteen, *The Shaping of Rationality: Toward Interdisciplinarity in Theology and Science* (Grand Rapids, Mich.: Eerdmans, 1999). The other two books by van Huyssteen quoted in this section are: *Essays in Postfoundationalist Theology* (Grand Rapids, Mich.: Eerdmans, 1997) and *Duet or Duel? Theology and Science in a Postmodern World* (Harrisburg, Pa.: Trinity Press International, 1998).

The First Alternative: Separate but Equal Realms

Stephen Jay Gould's *Rock of Ages: Science and Religion in the Fullness of Life* (1999) is a commonsense book espousing what many of us feel, namely, the correctness of the old cliché that science gets the age of rocks and religion the rock of ages; science studies how the heavens go, religion knows how to go to heaven. Gould argues for his principle of NOMA or non-overlapping magisteria. A magisterium is "a domain where one form of teaching holds the appropriate tools for meaningful discourse and resolution" (*Rock of Ages,* 5). The two distinct disciplines or magisteria are therefore each devoted to a central facet of human existence. Gould makes two primary claims for NOMA. First, these two domains hold equal worth in explaining the wholeness of human life. Second, they remain logically distinct and fully separate in their style of inquiry (58–59).

Gould, a Harvard professor and curator at Harvard University's Museum of Comparative Zoology, illustrates his argument for separate but equal domains with the troublesome and thorny issues of the evolution of human beings. He notes that in his *Humani Generis* (1950), Pope Pius XII defended the view that Catholics are free to entertain the hypothesis of the evolution of the human body under the magisterium of science so long as they also accept the divine creation and infusion of the soul as specified by the magisterium of the Church. Gould points out that Pope John Paul's statement in 1996 granted more authority, perhaps even equal authority, to science to the extent that "new knowledge has led to the recognition of the theory of evolution as more than a hypothesis" (*Rock of Ages,* 81). Gould agrees wholeheartedly: "The details of this contrast provide my favorite example of NOMA as used and developed by a religious leader not generally viewed as representing a vanguard of conciliation within his magisterium" (*Rock of Ages,* 75).

Stephen Gould's example is like believing the world is flat because we stand upright on its surface. It is neither good science nor good theology to let people believe the truth is just sensible. Likewise, it is neither good science nor good theology to let people believe that the soul is immediately created by God and infused into an evolved human body. Both make sense, but good science and good theology are usually much more complex. Gould's example is one of those perennial questions that illustrates the weakness of letting two distinct teachings each go its own way. God creates the soul and evolution creates the body, but to let each stand on its own excuses us from asking a number of questions. How do these two very different entities come together? Do we abandon what we have learned from the social sciences concerning the psychic-somatic fusion and the mutual development of body, mind, and spirit over time (not instantaneously)? How do we so easily separate body and soul when the soul is described by Arnold Kenseth as the "condition of being ready for God in the hour of his coming" or as "God made flesh, heaven clamped to the bones" (*Sabbaths, Sacraments, and Season,* 23)?

Unfortunately, Gould does not seem to be aware of the broad-based interest in reshaping epistemology as a more holistic process. Theoretical physicist and Nobel laureate Murray Gellmann express the predominant sentiment when he writes:

> When dealing with any non-linear system, especially a complex one, you can't just think in terms of parts or aspects and just add things up and say that the behavior of this and the behavior of that, added together, makes the whole thing. With a complex non-linear system you have to break it up into pieces and then study each aspect, and then study the very strong interaction between them all. Only this way can you describe the whole system.[5]

If anything qualifies as a nonlinear, complex "thing," it is a human being. Thomas L. Friedman, who writes about the phenomenon of holism in the domain of economics, comments on

5. Quoted in Thomas L. Friedman, *The Lexus and the Olive Tree* (New York: Farrar, Straus & Giroux, 1999), 24.

this quote by Gellmann saying: "We have to learn not only to have specialists but also people whose specialty is to spot the strong interactions and the entanglements of the different dimensions, and then take a crude look at the whole" (*The Lexus and the Olive Tree*, 24). Most theologians and scientists are content and qualified to look at the particular parts and practice their craft within their own magisterium, but for those few who have the cross-disciplinary training to spot the strong interactions and entanglements, NOMA is an unnecessary barrier.

Gould would have us believe that we can produce a *pure* science and a *pure* theology, each queen of its own domain. I see this as the underlying but unspoken assumption for his claim for non-overlapping subject matter and his insistence that each must ask different and logically distinct questions. It may very well be that scientists and theologians practice different reasoning strategies and have different governing interests but as a principle, NOMA bars the door to interdisciplinary inquiry into questions that resist reductionism and call for some form of narrative truth. We might say that Gould is philosophically naïve, because he does not appreciate the way any "truth" is embedded with its own particular context or paradigm or tradition. Because of this inevitable network of epistemological entanglement, or MacIntyre's "language-in-use," there is no surgical procedure to separate the domain of "is" (science) from the world of "ought" (religion). We can sympathize with Gould when he reminds us of the many misguided misuses of science when "scientific fact" became a tool for oppression. This, Gould would have us believe, invariably happens when we break one of NOMA's chief principles that "factual truth, however constituted, cannot dictate, or even imply, moral truth." Gould, however, is trying to recreate a world of clear and distinct ideas. As postmodern historians and philosophers have taught us, science operates with a methodology that is consciously and unconsciously social and political. We are quick to recognize how various forms of Darwinism fed the Nazi psyche for domination and a superior race, but forget that Newton's physics of absolutes engineered a hard, mechanistic, materialistic worldview.[6] The "is" of science is always an interpreted or contextualized "truth," which is all the more reason that factual truth be countered with verbal truth and that all meta-like narratives claiming to know the world as it is be opposed by a different viewpoint. We should call upon philosophers, not scientists or theologians, to help expose the way we have mixed, harmonized, and conflated categories and assumptions.[7]

There are times and places where science and theology should be allowed to maintain a critical distance in order to protect the integrity of each. When Langdon Gilkey testified against the scientific creationists at Little Rock in 1981, he used the "two language" argument to help convince Judge Overton that evolution speaks the language of science while creation speaks the language of religion, and the two should be separated in public education.[8] This, however, is not Gilkey's position as a liberal theologian outside of public education, because he would argue that both religion and science illuminate the same reality but from different perspectives.[9] The two-language argument is a valid approach, therefore, insofar as it serves to preserve the distinctive features of science and religion and to discourage simplistic answers to complex questions. When there is a healthy separation, it is possible to ask rather than assume, for example, whether the biblical writers intended to provide scientific answers to scientific

---

6. For an interesting read of the Newtonian synthesis as a "male thing," see Carolyn Merchant, *The Death of Nature* (San Francisco: Harper & Row, 1983), especially chaps. 7 to 10.

7. See Holmes Rolston III, *Genes, Genesis and God* (Cambridge: Cambridge University Press, 1999), where he unpacks the hidden philosophical assumptions of Richard Dawkins's selfish gene (47–50), and his "gene law of universal ruthless selfishness" (81–86), and the way such words as "altruism" and "selfishness" are borrowed from culture and are forced to fit a biological world that is amoral (277–80). In *Mystery of Mysteries* (Cambridge, Mass.: Harvard University Press, 1999), Michael Ruse makes the valuable distinction between epistemic values and nonepistemic values, the former providing standards for scientific objectivity and the latter meeting sociopolitical needs. He concludes that science is a product of both and is therefore both objective and subjective.

8. Langdon Gilkey, *Creationism on Trial: God and Evolution at Little Rock* (Minneapolis: Winston, 1985).

9. See his *Nature, Reality, and the Sacred: The Nexus of Science and Religion* (1993).

questions. The separation of church and state may require public school teachers to separate the language of science from the language of religion, but the ultimate goal is an interdisciplinary approach that proceeds to the next level.

Gould's fear is that if you mix or harmonize the two magisteria, you will end up with bad science and bad religion. Gould's usual insights into history fail him at this point because NOMA as separate and methodologically distinct magisteria was never the dominant principle during the ascendancy of science as the new dominant paradigm. Gould may be correct that Newton, for example, "spent far more time working on his exegesis of the prophecies of Daniel and John, and on his attempt to integrate biblical chronology with the histories of other ancient peoples, than he ever devoted to physics" (*Rock of Ages,* 84). Newton, however, is representative of scientifically minded individuals who mixed science and religion because they believed that together they spoke a single truth.[10] Until science became professionalized, there was no felt need to pit science against religion. As long as science ruled as queen, a principle of equal but separate domains was impossible. As rival paradigms emerged and methodological questions were raised and decided, the rivals separated. If my historical analysis is legitimate, in contrast to Gould's, then science and theology (theology as distinct from religion) were siblings squabbling over the same territory. What we see in Charles Darwin and Thomas Henry Huxley, both cited by Gould, is their demand for methodological consistency and the separation of domains. Whether this pattern should be continued is worthy of debate. J. Wentzel van Huyssteen, for example, argues that even if theology and science are about different domains, "we could never remain content with a nonfoundationalist pluralism of unrelated interpretations."[11]

---

10. Dobbs and Jacob, *Newton and the Culture of Newtonianism,* 8–12, 20–31, 53.

11. Van Huyssteen also quotes from Ian Barbour's 1990 book, *Religion in an Age of Science:* "If we seek a coherent interpretation of all experience, we cannot avoid the search for a unified worldview" (van Huyssteen, *The Shaping of Rationality,* 184).

NOMA is a sensible starting place but science and theology have matured sufficiently to navigate a more complex interaction where they intentionally map out conundrums of importance that require the wisdom and experience of both disciplines.

The Second Alternative: Convergence

Nancey Murphy's *Theology in the Age of Scientific Reasoning* (1990) is a prime example of a convergence approach from a postmodern perspective. The same might be said of van Huyssteen's culminating work, *The Shaping of Rationality* (1999), except that Murphy seeks more than just a shared meeting ground. Both Murphy and van Huyssteen want very much to end theology's epistemic isolation and to do so by accepting certain postmodern conditions. From my perspective, Murphy's *Theology in the Age of Scientific Reasoning* looks like one more attempt to force theology into a scientific mold.

The argument goes that theology will have a viable position only if it has a respected methodology; this means the acknowledgement of certain postmodern suppositions. They include: (1) a change from foundationalism to holism in epistemology; (2) a change from the modern emphasis on reference and representation to an emphasis on language as action, and "meaning as use; (3) a change from individualism to the irreducible importance of communal thinking and consensus."[12]

---

12. As summarized by van Huyssteen, *Shaping of Rationality,* 97–98. In drawing his arguments to a conclusion in *Essays in Postfoundationalist Theology,* van Huyssteen gives his list of postmodern assumptions theologians and scientists should accept.

1. "the crucial role of being a rational agent, and of having to make the best possible judgments with a specific context, and with and for a specific community.
2. the epistemological fallibilism implied by contextual decision making;
3. the experiential and interpretative dimension of all our knowledge;
4. the fact, therefore, that neither science nor theology can ever have demonstrable certain foundations" (*Essays,* 264).

I find myself accepting the first presuppositions with some reservation. Murphy would have us go the next step by reconstructing theology as an empirical research program. It is the second presupposition that concerns me the most. Her approach, like that of almost every theologian who wants to respond positively to the postmodern challenge, is to forget ontology and fixate on epistemology. In turn I would ask: To what extent can there be methodological convergence if the reality, the ontological other, is bracketed as not important or not relevant?

The approach here is not convergence in the usual sense of seeking a unified or holistic worldview where different pieces are fitted together to complete the picture, although at times it can be that. The approach is instead consonance, in the sense of different harmonics coming together to make a complete musical tone. In *On the Moral Nature of the Universe* (with George Ellis, 1996), Nancey Murphy does not adjust her methodological stance but allows theology the highest level of understanding in that "certain aspects of reality require the context of a vision of the purpose of the whole in order to be fully intelligible" (220). Reality is indeed such that no single methodology is sufficient; therefore, I wonder why we would want methodological convergence.

Holism as an epistemological model is an attractive option for advocates of this second alternative, because it presumes that both science and theology are confronted with the same postmodern challenge—how to adjudicate between competing truth claims without resorting to foundational claims. Murphy's type of holism does not believe that language has been totally discredited, only that it is limited because its meaning arises from its usage and from what it refers to. "A major tenet of holist epistemology," she writes, "is that philosophical knowledge is different from scientific knowledge only in degree—that is, only as a matter of its relative distance from experience."[13] What she

13. Nancey Murphy, "Postmodern Apologetics: Or Why Theologians Must Pay Attention to Science," in *Religion and Science: History, Method, Dialogue,* ed. W. Mark Richardson and Wesley J. Wildman (New York: Routledge, 1996), 112.

has in mind is the way beliefs about reality are always formed in webs of interconnected units. In holism, language functions identically for theology and science: as useful constructs and as useful not in the sense of depicting the nature of things but as in making the best argument and the better description of Christian practice, meeting the criteria of consistency and coherence (Murphy, *Anglo-American Postmodernity,* 117, 123). Epistemological holism, therefore, is the keystone for this second alternative.

In order for science and theology to collaborate, Nancey Murphy is inclined to treat the data of religious experience as objective phenomena, but "objective" only in the postmodern sense that scientific and theological "facts" are inevitably theory-laden and language-formed. The data of religious experience is fitted into a research program with hard-core and auxiliary hypotheses capable of being progressively tested.[14] Murphy acknowledges that in comparison with scientific data, the data of religious experience will not be of the same quality, reliability, or replicability (*Age of Scientific Thinking,* 173). This is precisely the weakness of the approach, because the experience and explanation of God is of a different texture than the experience and explanation of particles and quasars. The data of religion is highly personal and experiential. Theology reflects upon this data as it has been handed down (tradition) and written down (Scripture). David Tracy thus describes the methodology of theology as one of retrieval (*Plurality and Ambiguity,* 111–12). Murphy is confronted also by the long-standing problem of how to distinguish between religious experiences that represent encounters with God and those that do not. If theology is as methodologically convergent with science as Murphy contends, then this problem would diminish over time. But it has not and theology is forced to resort to communal

14. For a persuasive critique of Murphy in this respect, see Herrmann, "A Pragmatic Approach to Religion and Science," *Rethinking Theology and Science,* 12; and van Huyssteen, *Essays,* 79–83, 89–90.

consensus with the consequence of many voices and many authorities. Murphy's central argument that theology operates with empirical-like methodology is strained from the beginning. She succeeds only in showing how difficult it is to accommodate both scientific and theological explanations of reality into a common, holistic, postmodern epistemology.

The Third Alternative: Intersection

Like Nancey Murphy, van Huyssteen is primarily concerned with rescuing theology from its epistemic isolation by way of rationality. It is rationality, he insists, that has been the prime focus of the postmodern challenge. He endorses a holistic epistemology that opens a space for true interdisciplinary reflection. On the one hand, van Huyssteen should be appreciated as hammering the last nail into the stubborn notion that religion is subjective and science is objective, that theology represents a postmodern throwback while science is the truly modern form of knowing (*Shaping*, 27). In a way, he makes too much of the values that are shared between theology and science in the sense that he feels he must argue and demonstrate that theology is a rational discipline operating at the same level as science. Scarcely anyone in the academic world is still skeptical that theology and science do not "share the rich resources of human rationality." All this strikes me as just a different version of the younger sibling still trying to measure up. On the other hand, no argument can be too persuasive, because van Huyssteen takes as seriously as any theologian the creeping disease of fideism. If theology is going to be a dialogical partner, van Huyssteen argues, then it must allow *all* of its claims to be subject to critical review. Van Huyssteen is relentless in his attack on any form of foundationalism and sees fideism as the greatest temptation waiting at the door (*Shaping*, 108). Fideistic response takes many forms: when faith is allowed to provide the evidence for a rational argument, the Bible is presumed to be self-authenticating, inspiration functions as a form of intuitive knowledge, and revelation supplies a propositional foundation.[15]

Van Huyssteen places himself at the forefront of responding to the postmodern challenge of refiguring our notion of rationality, steering a course between modernism and postmodernism. The former is not an option because an Enlightenment mode of truth is wedded to disengagement, objectification, universal standards of judgment, oppressive meta-histories, and a correspondence model of truth. Postmodernism, on the other hand, quickly degenerates into unmitigated pluralism. This "splitting the difference" between modernity and postmodernity, as van Huyssteen develops it, means defending a position of managed dissent and accepting the rationality of a healthy pluralism and "the fact that diversity can play a highly constructive role in human affairs" (*Shaping,* 272). Rationality, then, is not primarily about consensus building, because it may impede the differing rational reasons of a healthy pluralism.[16]

In another form of splitting the difference, van Huyssteen argues that a theological way of knowing differs from a scientific way "only in degree and emphasis" and yet the "two are very different reasoning strategies," with two sets of very different claims to knowledge.[17] It is clear that van Huyssteen shows a split personality concerning these critical issues. Whether he is successful—or if it is even possible to maintain both poles—deserves further discussion. We do know that van Huyssteen's sympathy lies with finding a common methodological meeting place, but the price he pays is to reduce methodology to rationality. Cer-

---

15. Van Huyssteen does not mention the historical Jesus debate but it is an ever-present issue. See Marcus Borg's accusation of N. T. Wright's position that faith is a legitimate reason for making historical criticism. M. Borg and N. T. Wright, *The Meaning of Jesus: Two Versions* (New York: HarperCollins, 1989), 232–33.

16. Van Huyssteen differs substantially with Murphy regarding the role of consensus and her argument that it is one of the hallmarks of a holistic epistemology. Van Huyssteen at times echoes my own position: "What theology needs in its discussion with a secularized scientific culture, however, is to show that what really challenges the shaping of rationality in postmodern theology is its ability to represent *an authentic Christian voice* in precisely a radically pluralist context" (99; emphasis added). See van Huyssteen, *Shaping,* 98–101.

17. Van Huyssteen, *Shaping,* 13; *Duet,* 9 and 41; *Essays,* 239.

tainly, the case has been made that science and theology share a pool of resources of appropriate evidence, provide good reasons, and make responsible judgments, but do they therefore necessarily function with the same kind of evidence, give the same kind of reasons, and depend upon the same kind of responsible judgments? Van Huyssteen strongly asserts that even though "theology and the sciences have different epistemic scopes, different experiential resources, and different heuristic structures," this does not mean "they also have different rationalities" (*Shaping*, 202). Consequently van Huyssteen struggles, as does Murphy, to convince us that although there are important differences between the way theology and science fit together epistemology and ontology, they are nevertheless methodologically convergent. Once you have acknowledged "the often radical differences in epistemological focus and evidential grounds," then I believe you are committed to answer why science and theology are not incommensurate (*Essays*, 264). Van Huyssteen cannot pursue this trajectory because it might undermine his overriding need to map out a postmodern rationality acceptable to both theologians and scientists.

It is a difficult position to maintain but it must be done; that is, to assert that there are rational standards that are transcultural and cross disciplinary *and* that each research tradition has a rational integrity of its own. This puts me into van Huyssteen's camp, but I cannot stay here. While van Huyssteen describes the relationship as a "graceful duet," this model reveals the inherent inclination in his position to round off the sharp edges. In deciding against separation and convergence in favor of "a wide reflective equilibrium" of various voices, van Huyssteen is a powerful voice for interdisciplinary dialogue.[18] I wonder, however, if seeking a wide reflective equilibrium suggests a certain

18. The phrase "a wide reflective equilibrium by various scholars" is van Huyssteen's conclusion to how we find "the safe but fragile public space we have been searching for: a space for shuttling back and forth between deep personal convictions and the principles resulting from responsible interpersonal judgments." See *Shaping*, 277–78.

accommodation to science. To say, as van Huyssteen does, that theology and the sciences have different epistemic scopes, different experiential resources, and different heuristic structures, is tantamount to depicting them as sibling rivals, albeit rational rivals. Van Huyssteen does little to account for inspiration and sin as indispensable epistemic components and revelation as an ontological "event." He acknowledges the necessity for dialogical partners to keep their belief commitments, but by omitting faith as a complement to reason, he glosses over what makes theological methodology unique.

Conclusion

Since van Huyssteen respects scientists as realists, he expects theologians to be realistic and to make claims that are explanatory and not simply descriptive. Van Huyssteen mentions, not hesitantly though sparingly, the importance of a common ontology of the observable universe. He writes, for example: "But if theology and the sciences share the rich resources of human rationality, then it is to be expected that there will be important parallels between the role of explanation in the sciences and in theology" (*Shaping,* 259). Van Huyssteen continues by highlighting the unique aspects of theological explanations. They are normally all-encompassing, deeply personal and often arise from vague and elusive questions concerning the meaning of life. What is disappointing is his typical threefold criterion for an adequate postmodern rationality: coherence, consistency, and optimal understanding (*Sharing,* 278). And he says this after his strong critique of Murphy's position of an internal realism.

> In the extreme form of this view, religious beliefs have no need for explanatory support and can in the end be seen as just part of a groundless language game. However, when theological beliefs become a species of belief whose truth is discovered only by means of criteria internal to language

> game itself, this leads not only to a relativistic understanding of justification, truth, and knowledge, but to an epistemological relativism that would be fatal for the cognitive claims of theological statement—precisely in an age of postmodern scientific reasoning. (*Essays,* 89)

Van Huyssteen is more of a critical realist than Nancey Murphy; he is critical of her conspicuous dash from theology engaged with the real universe to a performative language model. But his postmodern rationality, like Murphy's, is one that must not be tainted in any way by notions of truth as an ideal, or universal claims, or objectivity, or demonstrable certain foundations. He may think this brings his form of rationality into line with a postmodern science, but I have my doubts.[19] It also does not seem important to him that there are many forms of truth and that some are directed at the observable and some are intended to evoke in the human heart a response to a reality that is not observable. Van Huyssteen has not found a place or a reason to include within his epistemological model the phenomenon of discovery and resistance (the otherness of reality). If he had, then I would believe he is the realist he claims to be. In the final analysis, his realism, as for most theologians, becomes an ontology of no consequence.

We might ask how this third alternative differs from my position that theology and science are sibling rivals. First, van Huyssteen is not willing to assert that theology has a distinctive methodology to the degree that it stands as a methodological challenge to science. I believe van Huyssteen has succumbed to the postmodern deconstruction of ontology and therefore is vague about the way ontology and epistemology might fit together. He does not fully appreciate that dispersed throughout theology is a soft ontology coupled with an even softer epistemology resulting

19. Doubts arise in part because I am not convinced that those doing science share the postmodern philosophical rejection of universal claims, objectivity, and demonstrable certain foundations.

in a hesitancy to embrace theology's true vitality as guardian of verbal truth. In my judgment, van Huyssteen is a critical realist who seems to doubt that language has much referential or representative value with the ability to describe truthfully a universe with its own universal properties. His is a weak form of critical realism, while mine is a strong form.

It is both characteristic and ironic that Murphy and van Huyssteen are only interested in the epistemic implications of realism and make this the basis for the conversation between science and theology. I find it very perplexing that Murphy engages science by rejecting critical realism as a meaningful epistemological model. Van Huyssteen, Murphy, and I begin with the same postmodern assumptions,[20] but I differ substantially with them about the use of language to capture the significance of what is ontologically other and real. Inherent in the weakness of Murphy's approach is her dependence upon postmodern strains of philosophy, especially those philosophies of science that deprecate ontology and realism. Van Huyssteen, on the other hand, keenly explores other implications for critical realism but they are for the most part epistemological. My contention and concern have not been that methodological questions are unimportant, but the degree to which the conversation has been tilted in their direction.

– SECTION 10 –
THE CHALLENGE OF DIFFERENT FITS

Previously I have referred to the fit between epistemology and ontology and suggested that each discipline or research tradition has a particular fit. In the context of this section, it is not important whether science and theology are considered to be research programs (Imre Lakatos), research traditions (Larry Laudan), traditions (Alasdair MacIntyre) or epistemological-ontological fits, because they are all complex and comprehensive frameworks that

20. See n. 12.

include networks of interlocking pieces, governing interests, and faith dispositions.[21]

The Problem of Incommensurability

The backdrop of this discussion is the troubling quandary posed by postmodern philosophy. If we begin with a postmodern epistemology, as I believe we must, which asserts there is no uninterpreted access to reality and that experience is always contextual, then there are no absolute, interdisciplinary criteria to judge between competing frameworks of ideas. Van Huyssteen, who embraces such a postmodern epistemology, is led to ask, "Are we ultimately and fideistically, the prisoners of our research traditions and commitments? And if not, why do we choose to commit ourselves—often passionately—to only certain traditions, theories, viewpoints" (*Essays in Postfoundationalist Theology*, 33)? In other words, Kant's transcendental idealism and any such transcendental methodology is ruled inadmissible, because it assumes that there is a way to stand outside of one's context in order to choose one paradigm as the correct one.[22] Therefore, the question before us is whether there is a transcendent perspective capable of adjudicating truth claims, especially those of the larger type such as a tradition or an epistemological-ontological fit. I am making an important distinction between a transcendental method, which is impossible given postmodern conditions, and a transcendent perspective, which is possible. This section will defend the second position.

Few philosophers have been more devoted to this question of

---

21. See Nicholas Wolterstorff for a discussion of the interplay between governing interests and faith dispositions. Wolterstorff, *Until Justice and Peace Embrace* (Grand Rapids, Mich: Eerdmans, 1983), 168–71.

22. See Robert D'Amico's conclusion that Kant's hope for a transcendental philosophy was based upon the distinction between the "a priori" (before experience) and the "a posteriori" (experience itself), a distinction that "fails to hold if the only criteria for such demarcation are relative measures such as practical success, instrumental control, or meaningful coherence." D'Amico, *Historicism and Knowledge* (New York: Routledge, 1989), 143.

commensurability than Alasdair MacIntyre.[23] MacIntyre poses the question of incommensurability by asking what happens when there is an epistemological crisis within or between traditions. By tradition he means a body of thought that employs only its own standards of justification. Thus, when it comes to adjudicating truth claims, every tradition will appear to be justified based on its own standards of assessment.[24] This is to be expected. Jeffrey Stout has further characterized the authority of these fits as self-authenticating (*The Flight from Authority,* chap. 2). The question of incommensurability between traditions arises in many ways: claims and language that are untranslatable, different positions starting from conflicting basic premises, different conceptions of rational acceptability, radically disparate governing interests, and incongruent heuristic structures. Consider, for instance, the educational methods of science and theology. The education of a doctor of philosophy in religion is quite different from that of a doctor of philosophy in evolutionary biology. Both qualify as rational traditions but have radically different governing interests. My freedom of choice about what swath of experience I will do my reasoning about is severely circumscribed by the methods I have been taught. I might spend the next year reasoning about the aesthetics of a particular religion or the principles in nature that govern emergent behavior; in making the choice I am constrained by the methods of my discipline. Even if theologians and scientists share certain epistemic values, as professionals in their traditions they will rely upon distinctive research models and unique uses of language.

MacIntyre provokes the question of commensurability by examining what happens when an epistemological crisis arises. This,

---

23. *After Virtue,* 2d ed. (Notre Dame, Ind.: University of Notre Dame Press, 1984); *Whose Justice? Which Rationality?* (Notre Dame, Ind.: University of Notre Dame Press, 1988); *Three Rival Versions of Moral Enquiry* (Notre Dame, Ind.: University of Notre Dame Press, 1990).

24. In bypassing this considerable body of work by MacIntyre, van Huyssteen does not do justice to the issue of incommensurability. For instance, how would van Huyssteen respond to MacIntyre's repeated assertion that rationalities are immanent in human practice and, therefore, there is not a postmodern-type of rationality that transcends disciplines?

it would seem, can happen in two ways: first, when two traditions or research rivals are sufficiently different that to adopt the other's methodology would be a substantial and unacceptable change; and second, when one discipline reinvents itself in order to remain viable. The dethroning of theology was an instance when science and theology could not both be viable when it came to truth claims about the physical universe. Theology could overcome the epistemological crisis posed by empiricism by either reinventing its methodology as empirical-like or as uniquely distinct. This issue is still not resolved. The second kind of epistemological crisis arises when one tradition acknowledges incoherencies and anomalies that it cannot account for. Process theology resolved the crisis brought about by evolution by elaborating a noninterventionist model of God, and some theologians have responded with a kenotic interpretation of God to resolve the apparent conflict between a universe that is the intentional work of God and yet is chaotic and meandering.[25]

Philosophers and historians of science provide numerous examples of what can happen when one paradigm or research tradition demonstrates a superior problem-solving capacity.[26] Assuming that rival traditions have a certain degree of commensurability, MacIntyre comes to this conclusion:

> The possibility to which every tradition is always open, as I have argued earlier, is that the time and place may

25. A kenotic theology brings together a universe that evolves not only by chance but by providence, not only by suffering but by redemptive suffering. The universe is created and redeemed by a God who is self-emptying. See for example Nancey Murphy and George F. R. Ellis, *On the Moral Nature of the Universe* (Minneapolis: Fortress, 1996), 208–11, and John Polkinghorne and Michael Welker, eds., *The Work of Love: Creation as Kenosis* (Grand Rapids, Mich.: Eerdmans, 2001); and John F. Haught, *God after Darwin* (Boulder, Colo.: Westview, 2000), 109–11.

26. Larry Laudan's *Progress and Its Problems: Towards a Theory of Scientific Growth* (Berkeley: University of California Press, 1977) is nearly a classic in this regard. Laudan, along with Popper and Lakatos, is responsible for the "standard reading" of history that accounts for "progress" as the resolution of problems, incoherencies, anomalies, and inconsistencies. They deny any actual progress resulting from representing the universe as it is ontologically using absolute standards. Thus, conceptual change can be rational without claiming to be absolute. This is MacIntyre's position as well. It is essentially an internal and historical understanding of rationality. See n. 28.

> come, when and where those who live their lives in and through the language-in-use which gives expression to it may encounter another alien tradition with its own very different language-in-use and may discover that . . . it provides a standpoint from which once they have acquired its language-in-use as a second first language, the limitations, incoherencies, and poverty of resources of their own beliefs can be identified, characterized, and explained in a way not possible from within their own tradition. (*Whose Justice? Which Rationality?* 387–88)

MacIntyre is arguing that an epistemological crisis is resolved by the construction of a new narrative that makes it possible to understand *both* how one could intelligibly have held the original beliefs *and* why a new way of seeing should be accepted as superior. What Galileo achieved, as an agent of transition, is "not only a new way of understanding the old nature" (the Ptolemaic system and Aristotelian physics, which had failed to come together in a coherent way), "but also and inseparably a new way of understanding the old science's way of understanding nature."[27] Thus, Galileo has a particular place in the history of science, because he introduced a new explanatory narrative that enabled "the work of all his predecessors to be evaluated by a common set of standards." In like manner, Einstein's schema of interpretation enables us to understand why Newtonian mechanics is superior to Aristotelian and why relativity is the superior ontology.

This works very well when traditions are able to reinvent themselves from within. Science and theology, though, are such multitask, comprehensive traditions that neither is going to sub-

27. MacIntyre, "Epistemological Crises, Dramatic Narrative, and the Philosophy of Science," in *Why Narrative?* ed. Stanley Hauerwas and L. Gregory Jones (Grand Rapids, Mich.: Eerdmans, 1989), 152–57. McIntyre thinks the superiority of one narrative over another is more than Kuhn's puzzle-solving interpretation. "Were it not for a concern for ontological truth, the nature of our demand for convergent relationship between all the sciences would be unintelligible" (156).

sume the other, but one could compel the other to reframe a significant piece of its map of reality. In some respects, MacIntyre offers an idealized scenario—the confrontation of two traditions, especially two that do not share a common language-in-use such as theology and science, does not usually result in a dramatic new reframing. Even where there is commensurability and a healthy self-awareness of one's governing interests and belief dispositions, fruitful dialogue is difficult. Can we understand MacIntyre as suggesting that transcendent criteria arise from time to time to adjudicate between rival traditions? I think not. MacIntyre argues that rational and "objective" judgments can be made between competing systems but not by applying absolute (transcendent) standards. His argument is for a historical perspective where one paradigm so challenges another that something must give, and what wins out is a paradigm or narrative that is "the best account anyone has been able to give so far."[28]

If there is a choice, should science and theology remain rivals or should they strive to find a common set of assessment standards? While I believe that theology and science share principles of rational acceptability, they do not employ in their practice the same standards of justification. This represents both a measure of incommensurability and a functional epistemological impasse. Conversation is certainly possible, but I do not see a willingness on the part of theologians or scientists to surrender important governing interests and faith dispositions. Nor should they, for that is what makes them intriguing rivals. MacIntyre also suggests that epistemological progress is made "in the construction and reconstruction of more adequate narratives and forms of narrative and that epistemological crises are occasions for such reconstruction" (*Why Narrative?* 142).

---

28. See Robert Stern for a discussion of MacIntyre's historical criteria for deciding between truth claims. Stern, "MacIntyre and Historicism," *After MacIntyre*, ed. John Horton and Susan Mendus (Notre Dame, Ind.: University of Notre Dame Press, 1994), 146–60.

When Different Fits Clash

What happens when two disciplines have discordant fits? Galileo, Newton, and Darwin forced theology to reinvent itself; however, the reinvention consisted of adapting to a new view of the world and finding a place for new paradigms. It was not insignificant, but the underlying methodological crisis had not been resolved. Consider, for instance, the epistemological-ontological fits that characterize each discipline.

Attention has already been drawn to the disparate epistemic styles: one is masculine (assertive, doing, penetrating, breaking apart), while the other is feminine (receptive, becoming, listening, nurturing). The argument that turns theology into a problem-solving, explanation-seeking discipline has always been awkward and forced. Incommensurability is alleviated somewhat when science is defined as the investigation of the universe and theology as the study of God's creation. The universe and creation represent a common ontology, but the difficulties do not therefore disappear.

There are many possible resources to draw upon to underscore a fit peculiar to theology. Christian philosopher Louis Dupré is especially well qualified as a crossover interpreter. Well acquainted with postmodern thought, Dupré grasps the problem with uncommon clarity.

> This much seems certain: a critical examination that on the basis of pure reason, independently of the religious experience proper, attempts to establish or disestablish "the truth of religion" must indeed result in distortion. Because modern philosophical theories of truth were developed for the purpose of securing foundation for scientific practice, their principles may appear unfit for evaluating the specific nature of religious truth. (*Religious Mystery and Rational Reflection*, 27–28)

Dupré highlights the essential reciprocity between the receptivity of the subject in coming-to-know and the self-revealing nature of God. There is no way theologians should bypass the ontological reference to God who is not an object but a subject who is self-asserting: "But, whatever the final conclusion may be, the religious act certainly displays a distinct quality in the passive attitude that the subject of this act adopts with respect to its object. That object [God] appears as providing its own meaning rather than receiving it from the meaning-giving subject" (*Religious Mystery,* 7).

It would be difficult to find another statement so contrary to the postmodern turn to the self-asserting individual. Since the postmodern mentality revolves around the self as the maker of meaning, theology resists becoming a convert. Dupré is not ignoring the "screen of interpretation" that penetrates every level of experience, but he will not use it as an excuse to reduce truth to a coherence theory. "Coherence easily turns to closedness," not a healthy position for theology.[29]

Van Huyssteen boldly asserts that he does not believe "a theological hypothesis should ever compete directly with a scientific hypothesis" (*Duet or Duel?* 125). Is this because they each depend upon incommensurate language practices and therefore do not meet on the same playing field? Or is it because their reasoning strategies are similar enough that a higher truth is always available? The latter. In van Huyssteen's own example of natural selection, he finds the hypothesis is one that adequately explains evolutionary change but is inadequate unless it is integrated into a more comprehensive theological view, one that includes what Arthur Peacocke identifies as a "propensity for increased complexity."[30] Incommensurate conflict is thus avoided by integrating

---

29. Dupré struggles to balance two claims: religious truth is exempt from "having to apply the same criteria that condition purely objective knowledge," and the imperative "to enter into dialogue with other realms and to render its claims compatible with the 'basic' interpretations of common sense and the physical sciences." See Dupré, *Religious Mystery and Rational Reflection* (Grand Rapids, Mich.: Eerdmans, 1998), 31.

30. Arthur Peacocke, *Theology for a Scientific Age: Being and Becoming—Natural, Divine, and Human* (London: SCM, 1993), 67. For a similar model where theology does not contradict science but *enhances* it, see Hans Weder, "Hope and Creation," in *The End of the World and the Ends of God: Science and Theology on Eschatology,* ed. John

a scientific hypothesis and theological insight in such a way that we do not have to choose between blind chance and divine providence. I agree that the supposed conflict in this instance is based less on methodology and more on two opposing worldviews. We have moved beyond a modernist duel, as van Huyssteen argues, where "'objective,' universal scientific claims were starkly contrasted to conflict with subjective, 'irrational' theological beliefs" resulting in a polarization of absolute claims (*Duet or Duel?* 2). It does not seem to me, however, that theologians have succeeded in convincing scientists that a hypothesis of divine providence can be squared with natural selection. At the level of meta-paradigms, there are good reasons to think their language-in-use represents a core disagreement.

If we accept the task of rescuing theology from its epistemic isolation, the task becomes doubly difficult when inspiration and revelation are added to the mix. Once the theologian begins to say that God attends to the bringing to expression of what is ontologically given, the cry goes out "unfair," because it appears that some trans-historical privilege is being exercised. And that is not far from what can happen. The proper response is to affirm that in religious experience there is something not self-impressed upon the inquiring subject.[31] True to Barth's insistence that there is no immediate access to the voice of God, Graham Ward argues that any insistence on the immediacy and directness of revelation compromises our understanding God as wholly other.[32] "Hearing" the Word of God or encountering the voice of God does not

---

Polkinghorne and Michael Welker (Harrisburg, Pa.: Trinity Press International, 2000), 191–96; and N. Murphy, *Anglo-American Postmodernity,* 66–80.

31. This a hotly debatable statement and one where believers and nonbelievers, theist and atheist, disagree. For an interesting explication of a religious experience, see Nicholas Wolterstorff's opening discussion of Augustine's conversion event. Wolterstorff, *Divine Discourse: Philosophical Reflections on the Claim That God Speaks* (Cambridge: Cambridge University Press, 1995), 1–8. The author has much wisdom to bring to the subject of how God speaks by making careful and new distinctions. Also, see John Hick, *Disputed Questions* (New Haven, Conn.: Yale University Press, 1993), chap. 2 ("Religious Experience: Its Nature and Validity").

32. Graham Ward, *Barth, Derrida and the Language of Theology* (Cambridge: Cambridge University Press, 1995), 80–96.

come in moments of pure intuition or inspiration; rather, hearing and encountering happens by way of interpretation and so we have to read the Bible *as* history or *as* the Word of God. This is not very different from Bohr's early warning not to think the world is accessible directly. There is one major difference. Bohr was looking at the problem from the methodological side. Barth's hermeneutical scalpel was poised to cut out anything that diminished God as subject who is wholly other. Revelation, per se, is never the content of faith or the pure experience of faith, because it exists between the two in the sense that revelation is the encounter that disrupts human existence by pushing the knower back to listening again. Revelation calls into question every hearing of God, because every hearing and speaking of God wants to possess the divine as objective speech or objective event. One must keep in mind that it is the ontological difference of God as other that continually reasserts the asymmetrical relationship between God and creature. Does this mean that we never know anything of what God discloses to us? No, because God has actually initiated a self-disclosure in Jesus Christ and elsewhere.[33] These are events where God can be seen and heard. There is an oral and written witness to those disclosing events.[34] There is a community and a tradition of interpretation that attends to those events. They are present for our listening and awakening but not to be objectified, because they exist as the vehicle *through which* God gets us to listen to questions we dare not, and cannot, ask of ourselves.

As a contrary fit, one of the gifts that theology can give to science is to reestablish the world as wholly other. Both the word of God and the universe are such that they continually wreck

33. Revelation is a kenotic act on behalf of God. Thus Ward writes: "Kenosis is the fundamental operation of the Trinity, but whilst it institutes the revelation, the giving of God Himself, it is also the content of revelation—the form of a servant which is the true likeness, the true nature of human beings" (*Barth, Derrida,* 158–59). The "elsewhere" is the divine disclosure of decisive events, the topic of section 13.

34. Walter Brueggemann's theology of the Old Testament reinstates the importance of witness and testimony. See his *A Theology of the Old Testament: Testimony, Dispute, Advocacy.*

every form of representation since transcendence itself cannot be represented. This is not to say that representation is rendered useless, but it is limited. Science does not have a tradition of the object as subject, nor should it. What it lacks and ought to have is what theologians have been continuously urging, a sense that the universe is beyond human grasp and explanation and that this constitutes its transcendent dimension. Within theology (both Christian and Jewish), there is also the tradition that the text, or the event, interprets the interpreter, because the text or the event supersedes the subject. What is interesting to consider here is the possible redress to the double crisis in epistemology and ontology. For both theology and science, methodology has overwhelmed ontology because the otherness of God and the universe have been rendered harmless. An appropriate fit for each would include a "dialectical tension" or feedback loops between the knower and the known, wherein each interprets and is interpreted by the other.

The Best Time of Our Lives

A transcendent perspective is possible but not as something arising from within one's own knowing tradition. The philosopher George Grant observes, "there is a pressing need to understand our technological destiny from principles more comprehensive than its own" (*Technology and Justice,* 34). His example is that computer technology is not as neutral as we would like to think. Although it is true that we are free to place the computer into whatever service we choose, how free are we when the science that created the computer was shaped by the same mentality that sets the standards of its right or wrong usage? Can we exercise self-transcendence when the instruments of technology and the standards of justice are bound together, both belonging to the same history of modern reason? The co-penetration of knowing and creating results in a new reality making it imperative that we pay attention to the methodological need for space and time to "stand back" in order to include a different perspective.[35] This would mean that an empirical rationality could only be criticized by a theological rationality and vice versa.

There is an acknowledged admission that while the Los Alamos scientists were working feverishly on the bomb, they had not given much thought to the worldview inherent in even one atomic bomb. Even more to the point, they did not stop to contemplate the creature who was creating this new thing. As the time approached when the "technically sweet" was becoming a massive destructive force, a few scientists began to consider a new reference point. Three different petitions were drawn up asking President Truman to reconsider carefully what we were about to do. It is more than a historical footnote that the only sustained protest came from outside the Los Alamos community, namely from Oak Ridge, Tennessee, and Chicago. The Szilard petition signed on July 17, 1945, by Leo Szilard and sixty-seven colleagues asked Truman not to decide on the bomb's use without seriously considering the moral responsibilities that were involved, but Truman and his Interim Committee were prevented from seeing this petition.[36] Secretary of War Henry Stimson and Colonel Leslie Groves, who was overseeing the project, succeeded in keeping the Szilard petition—as well as a poll taken by Arthur Compton and the Franck petition, a protest signed by seven physicists—from ever reaching Truman or the Interim Committee. It is doubtful these objections would have made any difference, for there was an unspoken, implicit agreement allowing the scientists to do "their" thinking and politicians to do what they do best. Colonel Groves saw himself as the exception, which gave him the right to decide what Truman would see or not see. Nuel Pharr Davis observes, "those in Washington who knew much about the bomb did not know much about the military situation. Those who knew much

35. A similar chord is struck by William Barrett, *The Illusion of Technique* (Garden City, N.Y.: Anchor, 1979), 243–48.

36. Nuel Pharr Davis, *Lawrence and Oppenheimer* (New York: DaCapo Press, 1986), chap. 8.

about the military situation did not know much about the bomb. President Truman did not know much about either" (*Lawrence and Oppenheimer,* 245). The result was a separation between science and public opinion that fostered an atmosphere of isolation that is now regarded as the enemy of peace and good science.[37]

The scientists at Los Alamos were having the time of their lives because they were "set free" to pursue knowledge at its fundamental level. Perhaps because he was not among the brilliant scientists at Los Alamos, Freeman Dyson could observe:

> The sin of the physicists at Los Alamos did not lie in their having built a lethal weapon. To have built the bomb, when their country was engaged in a desperate war against Hitler's Germany was morally justifiable. But they did not just build the bomb. They enjoyed building it. They had the best time of their lives while building it. That, I believe is what Oppy had in mind when he said they had sinned. (*Disturbing the Universe,* 52–53)

It would be expecting too much to think these scientists could distance themselves from their duty in the "national interest." The only voices to be heard were those of males doing masculine science. George Grant could not be more on target when he writes "brilliant scientists have laid before us an account of how things arc, and in that account nothing can be said about justice" (*Technology and Justice,* 60).

The scientists cloistered at Los Alamos were not atypical in a historical period when scientific and political issues were stringently separated. What was unthinkable would be to "mix" science and theology. That idea may have been literally unthinkable, since there were limited choices and all of them were

---

37. This very same issue came to the fore when Wen Ho Lee, a weapons scientist working at Los Alamos, was accused of spying for China. See the "My Turn" response by David Pines, who decries the political fallout to close the door to the openness that is the hallmark of international science, in "Why Science Can't Be Done in Isolation," *Time* (September 27, 1999): 11.

presumed to be bad science. Even many years after "the scientific and technological miracle at Los Alamos," Joseph O. Hirschfelder, a former group leader in both ordinance and in the theoretical divisions, is blind to the sin of myopia. With good intentions he quotes Pope Paul VI to the effect that these scientific discoveries must be used for the good of all humankind. This becomes the basis for him to hold firm to the belief that our future is secure if "we can produce a set of new scientific-technological miracles which are needed to solve our present problems."[38] To use a theological cliché, Hirschfelder, neither then nor now, has any "hermeneutic (principle) of suspicion" that asks why we should have any reason to believe a scientific fit will give us anything but more of the same unless it is radically challenged by another point of reference. Poignantly absent is a framework that includes an ontological understanding of sin that affirms the anteriority of evil and the omnipresence of overreaching.[39]

This is not the place to explore how a theological teaching of sin changes how we think about epistemology, but I will suggest one possibility. A long theological and philosophical tradition associates sin with the coming-to-know process. In the biblical account, sin is the desire to know both good and evil. I interpret the phrase "good and evil" (Gen. 3:5) to mean knowledge of every kind. Sin becomes the overreaching rooted in the desire to know all things. The philosophical tradition coming from the Enlightenment tended to regard the knowing process only for its potential for good. "A modern conscience is," if it is anything, "modern by pinning all its hopes on rational consciousness" (Tracy, *Plurality and Ambiguity,* 73). Science, the ever-optimistic sibling, never wavers in its belief that it will eventually get it right because its

38. His reflections are found in *Reminiscences of Los Alamos, 1943–45,* ed. Lawrence Badash, Joseph Hirschfelder, and Herbert Brodia (Dordrecht, Holland: D. Reidel, 1980), 67–88.

39. This phrasing recalls Paul Ricoeur's language for sin in *The Symbolism of Evil* (Boston: Beacon, 1967), 258. Ricoeur notes that in the biblical story the snake, as the symbol of evil, is present without explanation. He interprets this detail to say evil exists; it is present without explanation.

methodology of empiricism is self-correcting. To state it more precisely, the method is right but not always its application. This explains why the hope persists that it is possible to educate sin out of existence, because as we get the method right, sin becomes less and less a matter of any importance. We encounter this same notion in William Leiss.[40] "But if we face something more elusive and profound than error," David Tracy writes, "if we face systemic distortion, than another intellectual strategy is also called for."[41]

Sin therefore is not only what we do under the cover of night; it is what we do when we are enjoying the technically sweet. Our sin is not that we reach but that we overreach. The reason we reach out to new worlds is that we are self-transcending beings. The core reason we overreach is that are never content just to reach or transcend. Sin is more than getting it wrong some of the time. It is ontological, because the desire to know all things is never far from the insatiable desire to know.[42]

Conclusion

There is nothing inherently misplaced in looking for what is fundamental, universal, and unifying. Science and theology should pursue this endeavor from their own particular governing interests, but each discipline will see the world from its particular perspective. That is the nature of all epistemological-ontological

40. See W. Leiss, *The Domination of Nature,* 151–55 and my discussion of Leiss's resolution of the drive to dominate in section 5, "Science Overpromises and Overreaches." For all the attention paid to epistemology, Nancey Murphy and Wentzel van Huyssteen neglect to seriously consider how sin affects a theological understanding of coming-to-know. When sin is understood as ontological rather than merely epistemological, the second perennial question of who we are is profoundly changed.

41. David Tracy, *Plurality and Ambiguity,* 73. Tracy observes that error is the only problem for an optimistic account of reason. "For the autonomous, mature, coherent, modern self can surely handle error by finding better rational arguments or more reasonable conversation as it moves ineluctably forward into further enlightenment" (74). It follows within this construct that if the modern, self-assured self can handle error, then it can handle sin.

42. Wolfhart Pannenberg develops this same thought around our restlessness and openness. See his *What is Man?* (Philadelphia: Fortress, 1962), especially chap. 1 ("Openness to the World and Openness to God").

fits: they want to be self-justifying and self-contained. The fruitfulness of a conversation rests with the disruption of one's cherished assumptions and faith dispositions. I am at a loss to find a better reason for scientists and theologians to converse about the kind of universe we live in and the kind of creatures we are. The goal is not to be content with divergent perspectives or to press prematurely for a unified map of reality. We exercise our capacity for self-transcendence to the highest degree when we are impelled to stand outside our own particular tradition and ask if there is an ultimate grounding for the entire story.

Part Three

Working toward a Narrative Truth of the Universe

THE THIRD PART of this book is more theological and constructive in nature. It begins by defining the unique qualities of narrative truth. The most important thing about narrative truth is its ability to bring together different forms of truth: historical, empirical, and theological. Narrative truth is about event, plot, characters, and place and how their significance can be configured to tell a story—not just any story, but one that answers the perennial questions of who we are, where we are, and the nature of the universe.

The validity of narrative truth depends upon the incorporation of decisive events. These events occur in history, evolution, and the universe. They are a mixed bag; what they have in common is the capacity to overflow with a surplus of meaning, to change the course of history, evolution, and thc universe, as well as to galvanize other events into mcaningful patterns. Without decisive events, time itself would be flat and trivial. Because there are events that protrude and call attention to themselves, we can discern storylines, identify plots that unfold, and detect directionality.

Revelation deserves a fresh consideration when it is viewed as a decisive event with the potency to galvanize other events and to generate a storyline that flows from the meanings of those events. It is important not to identify revelation with a supernatural cause or even to search for a divine cause. What makes an event disclosive is its place within the narrative, for it is the narrative itself that is revelatory of God.

This book concludes with a frank overview of where the conversation stands between theology and science. That same frankness is carried forward when I unpack some long-standing issues to illustrate that these two historic siblings still have a rivalry going. My hope is that the discussion will model a new level of maturity, because we have entered an era in which decisions about the future will never be so critical.

Chapter 5

The Elements of Narrative Truth

Tell me the landscape in which you live and I will tell you who you are.

—JOSE ORTEGA Y GASSET, quoted from Belden C. Lane, "Fierce Landscapes and the Indifference of God," *Christian Century* (October 11, 1989): 907

Working in prisons, I get to see the results of "bad" karma up close, although it's hardly any different outside the prison walls. Every inmate has a story of one thing leading to another. After all, that's what stories are.

—JOHN KABAT-ZINN, *Wherever You Go, There You Are,* 222

– SECTION 11 –
NARRATIVE TRUTH

What Is Narrative Truth?

Narratives are the means by which we make a coherent fit or meaningful connection between epistemology and ontology. Because narrative truth is inherently relational and connective, it encourages bridge building between physical nature and human nature. Narrative truth provides a single account of what is real and defines our place in it. It allows us to speak about plot, characters, direction, purpose, and intention in a way that is minimally acceptable to scientists. In the broadest sense, narrative truth seeks to affirm that life on this planet, as part of the universe, has a moral coherence as well as a unity of purpose and meaning.

The many forms of narrative—drama, epic, history, myth, sacred story, ballad, opera, ritual reenactment, stained glass window, the novel, and so on—reflect the narrative quality of experience itself. Stephen Crites argues that consciousness is at least in some rudimentary sense narrative, because without it memory would have no coherence at all. Memory provides form in order to save us from "a disconnected succession of perceptions."[1] We call some stories sacred because they name those places, individuals, events, and plots that are essential to "orient the life of people through time." Sacred stories, as distinct from mundane narratives, are fundamental stories because they address the perennial question of who we are and the nature of the cosmos. They are not distinctive because gods are mentioned or celebrated but because a sense of self and world is created through them. Sacred stories are powerful because "the story itself creates a world of consciousness and the self that is oriented to it" (*Why Narrative?* 71). Residing at both conscious and unconscious levels, like a great reservoir, sacred stories cannot be told directly. They

1. Stephen Crites, "The Narrative Quality of Experience," in *Why Narrative?*, 65–88.

emerge into the light of day as memorable episodes and recollected images to be shaped over and over by contingencies of the moment (*Why Narrative?* 65–88).

Narrative truth is not so naïve as to think there are bits of truth waiting to be uncovered, whether in our investigation of the physical universe, in the recovery of history, or in the reconfiguration of a person's own life. Nor does it subscribe to a chronological model whereby once you have the order correct, you have truth. The epistemological process is not akin to opening a box but to writing a book or composing a song. Discoveries are made, revelatory moments occur, and decisive events bring forth new directions and meanings. From a psychological point of view, we could say that narrative truth can be defined as the criterion we use to decide when a certain experience has been captured to our satisfaction; it depends on continuity, closure, and the extent to which the fit of the pieces takes on an aesthetic finality. Narrative truth is what we have in mind when we say that such and such is a good story, that a given explanation is fruitful, that one solution to a mystery must be true. Once a given construction has acquired the form of narrative truth, it becomes just as true as any other kind of truth, because truth is always the interplay of construction, memory, and the real.

The reader may have the impression that there should be a single narrative *truth*, and while that is an ideal, we are only able to write narrative *truths*. The narrative truth of the universe is not a new worldview that tries to harmonize different kinds of truths. From the outset, I rejected a new worldview as naïve and impossible. Narrative truth seeks to incorporate little dramas as they make up a central narrative.[2] There is the short story as it is understood by Jews and Christians. For Jews, Scripture answers the leading question by saying that what God is up to in the world is covenant and Torah; for Christians, it is the inclusive

2. I find Walter Brueggemann helpful in this regard with his discussion of "little stories" versus the "great story." See Brueggemann, *Texts Under Negotiation* (Minneapolis: Fortress, 1993), chap. 3.

and unmerited love of God for the creation as it is particularly revealed in Jesus Christ. This is the short story. The long story tells about the birth of stars and planets, the emergence of life, the diversity of life forms, and the divine intention they reveal.

Narrative truth is a unique epistemological-ontological fit. It is respectful of the physically "real" and the conceptually "true" as distinct and necessary facets of learning what is valid about ourselves and our situation. It does not exclude the revelatory character of decisive "events" that overflow with meaning. Narrative truth postulates a progressive verification, because discovering truth or being inspired to transcend our given moment in time and place is more about hindsight than it is about intuition or experience. In other words, narrative truth sees—in the unfolding of matter, life, and history—a story that can be told where there is a progression toward completion, much like the completion of a jigsaw puzzle.

For the Sake of Conversation, Why Narrative Truth?

Both theology and science make truth claims about reality. What is not so obvious is that truth claims can be very different in their narrative form, whether it be history, a hierarchy of ideas, or a search for the origins of what exists. Those who think narratively take up what is past in order to proceed further and thereby provide us with a more complete account of who we are and of our place in the universe. As both theology and science have matured, they have assigned greater and greater significance to truth that is connected. Consider the claims that have been made for intuition, induction, deduction, discovery, logic, creativity, inspiration, or revelation. All of these have at one time been used to justify a truth claim, but it is decidedly false that any of them are isolated events of discovery or revelation. The crucial experiment, the inspired moment, the creative leap, the religious experience are indeed unique events, but what they mean and how they are to be put to use depends upon their connectedness. The truth behind narration is that someone, theologian or scientist, whether by inspiration or logic, believes that various pieces fit together in a certain way. The truth about stars is the story of how they began, developed, and will end. The truth about our species is likewise the story of its genesis, evolution, and will to intend something.

Truth as narration is essential for science. The strength of science is its intellectual ordering of sensory data that are translated into facts about the nature of the universe and knowledge of the real. But what are "facts" and what is "real"? These are questions as difficult as any asked by theologians. Are the answers to be found just by looking through an electron microscope or by examining the images on a photographic plate? For the numerous reasons recounted in this book, the reply is a resounding No. In a universe with a singular beginning and where everything is interconnected, there is a coherent narrative to recover. Does that history also take the shape of a story? It does to the extent that there is anything like drama, adventure, plot, characters, memorable events, or intention. Insofar as theology tells a story from the perspective of faith and history, science should listen to hear if there are echoes of that story in the unfolding of the cosmic adventure. Insofar as science unfolds the evolution of what exists independently of the human mind, theology should listen to hear if there are echoes of that story in its understanding of redemption.

Truth as narration is more than the history of evolution and survival. The narrative tells a story about reality. As Gabriel Fackre observes, "reality meets us in the concretions of time, place, and people, not in analytical discourse or mystical rumination"[3] We are inveterate storytellers because that is how we try to answer the perennial questions of who and why we are. Ian Hacking, following the lead of the Leakey family, suggests the origins of language lie not with the need to communicate

3. Gabriel Fackre, "Narrative Theology: An Overview," *Interpretation* 37 (October 1983): 341.

during hunting expeditions where nonverbal signals are critical, but with boredom as people gathered about campfires (*Representing and Intervening*, 135–37). In this scenario, language has more to do with recounting the hunt than the hunt itself and, perhaps, with impressing the opposite sex (a sophisticated form of mating call so universal in nature). Scientists and theologians concern themselves with different kinds of narrative truth—that is what constitutes their distinctiveness and makes them good conversational partners. Accident and design, chance and necessity, beginnings and endings, chaos and beauty are the material for scientific theories as well as for myths and novels. Various governing interests motivate the way we interpret and understand all things. If we do not attempt narrative, we spurn the very capacities we have as human beings.

Without narration, how can we begin to address such questions as what is matter, what is life, or how did the universe begin and how will it end? It profits us little to look at each event in itself. Chance and destiny, coincidence and providence are played out over long stretches of time; they only make any sense in hindsight. The intense interest science has recently shown for beginnings is more than just a passion to know what happened back *then*. Our fascination with the origins of *Homo sapiens* or the universe is also our interest in what it means for us, not just idle curiosity.

"The impulse to narration," as Amos Wilder calls it, is deeply rooted in our sense of time, continuity, and adventure.[4] At whatever level of development—awareness, thinking, reflection, deliberation—we map experience in order to orient ourselves in time and space. Individually, we tell stories to find ourselves; we listen to stories in order to find ourselves within the story. To grasp at antecedents and establish connections with what has gone before is our attempt to be prepared for what happens next. "The terror of history" (Eliade) and the terror of being "homeless in time and circumstance" (Wilder) are not far apart, since both

4. Amos Wilder, "Story and Story-World," *Interpretation* 37 (October 1983): 360–63.

terrors stem from a fate worse than death, namely, the monotony of ordinariness and the lack of adventure. Witness, for example, the use of boredom as a form of torture and the torture teenagers feel when they are being "bored to death."

Our personal stories are bound up with the world we inhabit. Ancient myths, whether Chinese, Semitic, Egyptian, or Greek, tell stories where cosmology is the subject of the narrative and not just the framework. When human beings are introduced, the stories become a threefold drama between the divine, mortals, and the cosmos. Common places become holy places because there is such a thing as "a spirit-of-the-place" that serves to help define who I am and who we are.[5] Consider the part played by holy places in the Old Testament. The Garden of Eden, Horeb, Hebron, Mount Sinai, Jerusalem, Mount Zion are more than places to locate on a map and they are more than landscape. They are as integral to the story as the characters themselves. Cosmology is always more than the backdrop or canopy. It is integral to the story, as science fiction writers (for example, Frank Herbert in his *Dune* series) understand better than most theologians and scientists.

When believers seek to know what God has done or what God is doing, the inquiry must include the unfolding of a series of events. That is why the Old Testament is primarily "salvation history" and the New Testament is primarily gospel stories. Even when it comes to those paradigmatic events like the crossing of the Sea of Reeds (Red Sea) or Christ's crucifixion, the faith of the believer versus the skepticism of the nonbeliever is not a decision about one event per se, but about its meaning within the larger story in which it is enfolded. The Egyptian pursuers and the Roman soldiers at the foot of the cross or, for that matter, the Hebrews and the disciples, did not necessarily see immediately an

5. Belden C. Lane, *Landscapes of the Sacred* (New York: Paulist, 1988); Calvin Luther Martin, *In the Spirit of the Earth* (Baltimore: Johns Hopkins University Press, 1992). Martin argues that the oldest hunter societies reveal "that words and artifice of specific place constitute humanity's primary instruments of self-location" (103). Both books are excellent arguments for the importance of place. See also the next section.

act of revelation in these events. That came later when the pieces began to fit together to tell a story.

Storytelling functions in the history of religion in precisely this way: to relate the individual person, tribe, or society, to a larger framework or "sacred canopy" as Peter Berger calls it. Through stories, in the form of myths, folk tales, legends, chronicles, biography, short story, or anecdote, the group finds a social identity. Lonnie D. Kliever writes:

> In other words, these stories serve the metaphysical function of linking the individual to the mystery of the universe as a whole, the cosmological function of furnishing an intelligible and heuristic image of nature, the sociological function of articulating and enforcing a specific social and moral order, and the psychological function of marking a pathway to guide the individual through the various stages of life. (*The Shattered Spectrum*, 153)

After a long history of improper focus, perhaps we are now able to recognize that the crucial difference between the profane and the sacred is not the distinction between natural and supernatural causations, but rather the difference in how we interpret the narrative results of these events. There is one notable difference. While we delight in the sheer play of fancy, like our prescientific predecessors, we require that we distinguish between fantasy and the real. This is why theology cannot do without science. Theology does not so much look to science for mathematical precision as it requires fantasy to be distinguished from fact.

Obstacles to Narrative Truth

Does narrative truth presume a correspondence between the "actual state of affairs" and their representations? Does the veracity of narrative truth rest with its ability to convince us about the accuracy of the correspondence between verbal truth and ontological reality? I am not ready to declare that such a fit is dead or passé, but it surely is not the same as when theology was queen. Beginning with the presumption that narrative truth is a unique vehicle for making connections between the ontologically real and the verbally true, we face two complaints. The first is that narrative truth is a weird hybrid and cannot be taken seriously. Second, it places fiction and nonfiction on equal footing.

The difficulty we face resembles asking the average person if fiction and nonfiction can refer to the same truth or if each speaks a truth of its own. The common impression is that one is false and the other true; one is real, the other is fantasy; one is objective, the other is personal. For those of that mindset, narrative truth, like story, has no epistemological validity because it is an artificial construction imposed by humans on the random sequence of events. However, any postmodern thinker knows the distinction between fiction and nonfiction masks a deeper reality. Both lines of criticisms overlook that human beings are continually reconfiguring experience into meaningful units. In many respects, the difference between fiction and nonfiction is the level of sophistication and self-consciousness with which the reconfiguring is done. The fact that scientists tend to do this differently than philosophers, or that I can do it personally and subjectively at one moment and impersonally and objectively at another moment, does not negate the obvious; namely, if we do not reconfigure experience by joining it with other experiences and other conceptualizations, it will be forgotten. This is true whether we are speaking of a life event, a historical event, a chemical event, a physical event, an experimental event, or an imaginative event. Let us not forget that all experience is already framed by previous experience, just as every event is framed by previous events. Holmes Rolston expresses what is by now almost the obvious: "One has to make up a story to catch and interpret the history, just as one has to make up a theory to catch and interpret the empirical facts. Recounted events are story-laden, but this makes them no less true than the theory-laden data of science" (*Science and Religion*, 277).

One can assert, as Amos Wilder does, that there is no "world for us" until we have named it, put it into words, and told a story about it. Talking about it has shaped what we take to be the nature of things. However, as Wilder points out, even if the only worlds we have are those that have been put into words, we concede too much to those language philosophers "who say there is no world *except* language world. They push too far their valid insight that stories evoke their own reality. The fact is that there is a prior sense for the real which pervades and tests all language and all stories."[6] The clue to what joins story worlds and real worlds is the motif of interest. Here we must take into account the wide difference in the *genres* of story. The range is as wide as fables, which seem to be inspired by sheer delight in the play of fancy, and scientific reports that want to connect cause with effect. But Wilder insists every story "is governed by the same urge to map our being and to know who we are and where we stand" ("Story and Story-World," 361). There is a common readiness to ask what happened, where did it all begin, and how will it turn out. The important question about fiction and nonfiction is not which one represents the "real" world, but what is the interest of the teller-writer. For the sake of truth, let us affirm "there is a prior sense for the real which pervades and tests all language and all stories" ("Story and Story-World," 362) *and* the freedom to map experience by the use of imagination and liturgy.

We are not far from the epistemological issue that has followed us from the beginning: Does reality have an order and meaning in and of itself? The argument I have pursued is that the intrinsic characteristics of both the physical universe and conceptual truths have a life of their own; where there is a proper fit, they convey a deeper truth. Fiction, therefore, "derealizes" or breaks the bondage of the real, thus opening up new historical possibilities (Peter Hodgson, *God in History,* 161). When one *acts* based on an imaginative variation, a new direction is brought into reality and history itself changes. When we effect change, that too

6. Wilder, "Story and Story-World," 362.

becomes a new history. To accomplish its aim, fiction must borrow from history just as history borrows from fiction. To put it in Hegelian terms, "the ideal must be riveted to the real, and the real to the ideal" (*God in History*, 161). When scientists intervene in reality, it is changed and again changes when new information becomes a new theory guiding the next intervention-observation.

Every narrative, as Paul Ricoeur reminds us, is not simply self-referential. We should not fall under the spell of the ontology slayers with their preoccupation with language as solely an internal system of signs, because that is not the intent of those who write or tell stories. Both fictional and nonfictional narratives propose a world we might inhabit, a vision of a world that has been reconfigured for some reason. But the reconfiguring is done differently: "fiction primarily by metaphorical reference, history primarily by reference to 'traces of the real,' " and science by abstracting what is universal (*God in History*, 92).[7] Yet, all who tell stories, interpret history, and discover laws do so "in order to enlarge, enrich, and reshape the world that will serve as the horizon for one's own engagement in a newly configured, transfigured praxis" (92).

The crux of the issue is whether we can accept both verbal truth, where the warrants or justifications are primarily how words are internally organized, and scientific truth, where the warrants are primarily referential and the ontological reference is indispensable. By accepting both, do we encourage the notion that theology and science have nothing much to say to each other, since they each have a proper domain? Yes, except that would mean returning to a separate but equal mentality. The postmodern understanding of science will not allow the clean separation for all the reasons I have continuously asserted. Scientists

7. In his "Afterword," Simon Schama laments the insoluble quandary of the historian: "how to take the broken, mutilated remains of something or someone from the 'enemy lines' of the documented past and restore it to life or give it a decent interment in our own time and place. . . . Historians are left forever chasing shadows, painfully aware of their inability ever to reconstruct a dead world in its completeness." See Schama, *Dead Certainties: Unwarranted Speculations* (New York: Knopf, 1991), 319, 320.

must work with forms of verbal truth as well as history; together they are always a mixture of event and reconfiguration. Theologians are finding more and more reasons to consider the universe in its givenness as essential for what they assert about the identity and intentions of God.

The Eclipse of Narration

On the one hand the Enlightenment marks the disappearance of the naïveté that history contains its own meaning. However, the Enlightenment, as William Willimon points out, was also blind to the world that its own scientific positivism was busy creating, so it "behaved as if it were possible to have a world constructed without story and without community, which, in an ironic sense, eventually became its own story and produced its own community—a community of people whose story is the claim that they are individuals who have not been formed by story" (*Peculiar Speech*, 52)

During the flight to objectivity, to use Susan Bordo's descriptive phrase, the language of mathematics and mechanics displaced narration as the language of meaning.[8] But in this process of displacement, time and space had to be impersonalized, universalized, and everywhere the same. "Subjective time," writes Bryan Appleyard, as "our own private sensation of duration was implicitly humbled and our modern obedience to objective, measured time was born" (*Understanding the Present,* 32). The lawfulness of time eliminated the uniqueness of events and thereby negated any direction to time's arrow. Stephen Gould tells how Charles Lyle (1797–1875), the renowned geologist, operated with a methodology which homogenized all causes so that it would be possible to understand the way the earth once *was* by the way

8. Walter Brueggemann summarizes Bordo's book (*The Flight to Objectivity*) by saying: "It is perhaps not too much in summarizing Bordo to say that a sense of the loss of cosmic mother led Descartes to fashion an impenetrable masculinity that nicely linked 'objectivity' and masculine power." See Brueggemann, *Text Under Negotiation,* 4.

the world *is* still today. His conclusion was that change is continuous, always the same, and leads nowhere.[9] The universe itself was becoming too complex, too immense, and too paradoxical to provide much meaning for the modern person. The very small and the very big grew further apart, the physical universe and the biological worlds were partitioned into separate domains, and the self as subject and the world as object became separate disciplines of study.

During the post-Enlightenment period, the dualism between subject and object, between mind and matter, became intolerable and was replaced with historicism and existentialism: two extreme reactions. One focused on recovering the facts of the past and the other on disregarding the "facts" of the past in order to accent the present moment. Biblical interpretation was strongly influenced by both existentialism and historicism. From the former we were led to read the Bible as existential stories of decision. From the latter, Scripture was read as depicting the mighty acts of God in history, intended to redeem the world. The alternatives were hardly satisfying, because an account could be true only if it either moved me to a moment of ethical decision or if it reported history accurately. In attempting to restore a sense of narrative truth to biblical interpretation, Hans Frei argued that the texts were meant originally to be read as realistic narratives and so they should be read that way by us (*The Eclipse of Biblical Narrative: A Study in the Eighteenth and Nineteenth Century Hermeneutics*, 1974). What was reality for the biblical writers, however, is a long way from what a science-dominated culture thinks of as reality. Therein lies a tremendous hermeneutical gap that Frei did not handle well.[10] To Frei's credit, he was instrumen-

---

9. Stephen Jay Gould explicates the two primary metaphors for time central to the intellectual and practical life of the Western world. Time's arrow captures that sense of time where events and history are the irreversible sequence of unique events, while in time's cycle, events have no meaning as distinct episodes and time itself is reversible. See Gould, *Time's Arrow, Time's Cycle* (Cambridge, Mass.: Harvard University Press, 1987), chap. 4 for his discussion of Lyle.

10. What the Bible deems as real is not what I can regard as real. Science has changed that forever for most of us. Its narratives do have the power, however, to open us up to

tal in reversing the disjointed effect of historical criticism where the meaning of the text was dependent upon a historical reconstruction. Frei's realistic interpretation was meant to be a middle course between defending the ostensibly historical truth of the texts, as if that would secure its truth, and detaching the meaning of text from its ontological referent.[11]

Conclusion

Narrative truth has a unique way of connecting epistemology and ontology. Narrative truth is ontologically minded but not in a literal or strictly historical sense. It does not want to lose the ontological fact that first gave rise to its being put into words. The temporal gap between perception and conception, with language as the interpreting mediator, means that our knowledge is never direct or unconfigured. When the framework is narrative, the space between experience and conceptualization is extended both necessarily and as an asset, because words are not merely passive windows to the world but also the lenses that bring the world into focus. At times language will explicate the meaning that belongs primarily to its ontological reference, while at other times it creates its own meaning. Truth, therefore, is not literally, plainly, or immediately present.

Because its scope is expansive, narrative truth will accept many forms of truths—not by blurring distinctions, but by honoring them. These four kinds of truth are narrative truths:

---

truths that would otherwise be denied because our understanding of what is real is limited by our own perspective. The biblical narrative offers alternative ways for me to be in the world. George Lindbeck in Frei-like fashion argues, "a scriptural framework is . . . able to absorb the universe. It supplies the interpretive framework within which believers seek to live their lives and understand reality" (*The Nature of Doctrine,* 117). I would ask: How can a prescientific world serve as the interpretative framework for Christians unless we bracket each and every point of conflict, as if to say what happened does not matter?

11. See Garrett Green for his discussion of how to handle historical events in texts that are not ostensibly historical. See Green, *Imagining God: Theology and the Religious Imagination,* cited in "Theology in Search," 135.

1. Scientific or empirical truths—mathematical equations and mathematical predictions
2. Historical truths—anecdotes, chronologies, reminiscences
3. Metaphorical or symbolic truths—parables, myths, rituals
4. Theological truths—narratives and doctrines

Strict scientific and historical standards have raised the bar, so to speak, against conflating different forms of narrative truth because of the fear that one will degrade the other, especially if one form is considered speculative, or ungrounded. It is an understandable fear—but the price we have paid has been too great. Narrative truth incorporates theological truths that are read off historical events and historical events that evoke rich metaphorical resonance. In their debate about the meaning of the historical Jesus, New Testament scholar Marcus Borg asks the rhetorical question of N. T. Wright: How much of the Gospels must be historical in order "to give substantial content to the claim that in Jesus we see what God is like and what a life full of God is like?"[12] This is the kind of question Borg would ask because he does not think "the metaphorical truth of the gospel narrative depends upon it also reporting a specific historical event" (234). Wright responds by saying, "this God has not left us to speculate, imagine, or project our own fantasies onto the screen of transcendence; this God instead, through self-revelation, has given us such knowledge as is possible and appropriate for us" (214). To restate the issue: It is not whether historical truth is more valid than metaphorical, or whether theological truth is more valid than empirical, or that metaphorical truth is more convincing when attached to a specific historical event. The issue is whether a verbal form of truth is more susceptible to speculation and projection. As far as a generalization can take us, I would have to say that truth as rhetoric or as the best argument—the

12. Marcus J. Borg and N. T. Wright, *The Meaning of Jesus: Two Versions*, 235.

purest form of verbal truth—runs the risk of becoming conjecture. On the other hand, historical and empirical "facts" run the risk of never adding up to anything significant.

Narrative truth has the reputation, especially in its postmodern form, of suspending ontological considerations and collapsing epistemology into semiotics. All symbolic designations become relative, and the force of language is not to pursue truth but to persuade by the best argument.[13] Words are unhinged from any natural correspondence with reality because there is no means of arbitrating between words themselves and what they point to. In my judgment, narrative truth does not take this form of skepticism but holds onto the ontological dimension of revelation (disclosure). It does matter, therefore, which words are used to bring to expression the ontological event or the intention of the writer. Garrett Green, who espouses a moderate position, writes, "In the case of narrative texts, their truth depends not on their ostensive historical reference but rather on their claim to exemplify an aspect of the real world" (*Imagining God*, 13). I would state it this way: In the case of narrative biblical texts, historical reference is not their primary objective, because their purpose is to arm us so that we might be in the real world as faithful Christians. This arming is done in a variety of ways: explosive metaphors, striking analogies, stories that turn our known world upside down (parables), the memorable-decisive moment, the theological reconfiguration to time, and facts revealing the mystery and intelligibility of the universe.

The critical difference between narrative truth and literal truth is the length and breadth of the perspective. The nature of narrative truth is to gather up and bring together in a meaningful way the entanglement of plot, character, and setting, as well as the historical events that generate meaning far and beyond their initial incidence. Although reality is constructed in and through

13. For example, Marcello Pera, *The Discourse of Science*, trans. Clarissa Botsford (Chicago: University of Chicago Press, 1994).

language, there are certain words, phrases, concepts, and symbols that better serve to reveal God. Verbal inspiration requires that some words be regarded as more directly given by God than others; therefore, these words are not only adequate, they are completely adequate. But individual words themselves do not have a right to this claim. The context counts, because the context creates the meaning, be it quite ordinary or quite extraordinary. God is the overarching referent that moves and invigorates the multiplex of employments. Each text is touched by God. Each individual word may not be inspired, but God is the subject and the reason for the text; thus, those words put together undergo an intensification of meaning.[14] It is not the words themselves but the service they perform; while the function remains the same, the words used from culture to culture will not be the same. Substituting other words, phrases, or metaphors becomes necessary in order to recapture the original intent. Whatever new words we propose, however, they cannot stand alone but must be judged by the narrative context in which they operate and generate.

The unique contribution of narrative truth is how truth is built up. Narrative truth contests the modern prejudice that separates truth and reality and insists that one kind of truth predominate. What Paul Ricoeur says in regard to the Gospel of Mark—"it is in narrating that he interprets the identity of Jesus"—could be said about narrative truth in regard to the universe: it is in narrating that we answer the perennial questions about who we are and what our place in the cosmos is (*Figuring the Sacred,* 185). When we focus on the larger picture, we are forced to emphasize the significant events, much as an individual does when he or she tells the story of his life. When perennial questions are asked, one attempts to probe for the largest interconnected units. This

14. Paul Ricoeur clarifies that what unites and differentiates Scripture is precisely the naming of God. The naming of God, Ricoeur argues, is first a narrative naming. See Ricoeur, *Figuring the Sacred,* 217–35.

compels narrative truth to provide rich texture for many forms of truth until a picture emerges that makes the universe intelligible and our existence meaningful.

– SECTION 12 –
PLOT, CHARACTERS, PLACE

A story or narrative consists of plot, character(s), and place. The plot turns on significant events and connects these events. The characters of a story enact the plot over time and place through conflict and adventure seeking some resolution. The place shapes the plot and the characters because "place is the house of our being" (Heidegger).

As human beings with the capacity for self-transcendence, we remember in order to make meaningful connections about the past and to anticipate what might happen in the future. We have a natural curiosity about what happened, especially if it was something unusual, marvelous, or unique. Along with this curiosity comes an interest in the sequence of events. We cannot help but inquire about the coincidence of events or the patterns of behaviors. Because we cannot change what was, we can be regretful about the past. Because we can change what will be, we are anxious about the future. As creatures living from the past into the future, we are driven by the impulse to reconfigure time and locate ourselves, to grasp at antecedents, and to prepare for the unforeseen. This is why we tell stories.

Generally, the elements of narrative truth correspond to the perennial questions discussed in this book. Plot addresses the question concerning our place in the universe, character encapsulates the kind of creatures we are, and place describes the kind of universe in which we live. For a storyline to unfold, characters interact with both place and events. More broadly speaking, plot provides meaning in the context of the characters' home, community, country, world, and universe.

Plot

Plot has both an epistemological and an ontological dimension. The latter is most broadly conceived as the universe continually emerging into increasingly intense forms of ordered novelty, beauty, and heightened consciousness (John Haught, *The Cosmic Adventure*). Everywhere there is the continual struggle between order and disorder, existence and extinction, creativity and destruction. This makes for a world of landscapes and horizons, mystery and strangeness, the unknown and the unexpected. An ontology of becoming has opened the way to a universe of stories, which stands in sharp contrast to the flat, adventureless universe of Newton. Holmes Rolston writes, for example, "One reason that evolution is a much richer and more welcome theory than was the former belief in the fixity of species is that it makes possible vaster depths of story" (*Science and Religion,* 275). Regardless of where one looks—nature, history, the physical universe—one encounters emergent steps, decisive moments, memorable events, saints and heroes, the upslope of the life processes, the way the past is gathered and incorporated into the future. Humans are not incidental to this story because they have a unique capacity and desire to fit their own stories into a cosmic story. In short, neither our world nor the universe is flat. With human beings in the mix—creatures who create and destroy, love and hate, embody the struggle between good and evil, quest to see new places, and in general will leave nothing as it is—the plot thickens. Human beings not only interpret reality as plot, they escalate it.

As thinking creatures, we do something unique with the ontological givenness of the universe. The epistemological reconfiguring of what goes on in an emergent universe is the second dimension. We find a variety of ways to reconfigure time by drawing together goals, motivation, interactions, circumstances, locality, sudden reversals, tragedy, and coincidences. The epistemological side of plot is this act of bringing to consciousness what is overlooked, undeveloped, and hidden from the objective eye. In *Figuring the Sacred*, Paul Ricoeur provides us with the useful word "emplotment," the art and ability to bring together the episodic character of history, life, and evolution "*from* a variety of events or incidents or, reciprocally, in order to make these events or incidents *into* a story" (239).[15]

Only a small part of reality is accessible to us at any one time and thus every reconfiguration of time, whether it be fictional or nonfictional, logical or historical, is necessary. Time and reality must be thematized or they are lost or forgotten and fade into the abyss of nothingness. The raw materials of evolution, human life, and historical events are sufficiently pliable and rich enough to suggest a number of ways to configure them. The material world is inanimate and composed of perfectly similar quantum, lending itself to measurement and quantifying. On the other hand, something substantial is added when the human mind superimposes a sense of beginning and ending. Science is committed to the belief that reality coheres according to the language of mathematics. Story time emphasizes just the opposite: the unique, the anomaly, the disruption, the unresolved that coheres around purpose and direction.

Stories, as a particular form of narration, are particularly useful in catching the connections between happenings and responses. Stories are indispensable for bringing intentionality to the forefront, because intentionality is not apparent in the chronological sequence of events or in a scientific description of cause and effect. A plot requires that some intentionality, however obvious or hidden, conscious or unconscious, be worked out. The creation narratives in Genesis, for instance, express a divine intentionality.

"Intentionality" is a loaded word for theologians as well as for scientists. For those who steadfastly reject natural theology,

15. See also Paul Ricoeur, *Time and Narrative*, trans. Kathleen McLaughlin and David Pellaue, 3 vols. (Chicago: University of Chicago Press, 1984, 1985, 1988). I have greatly simplified Ricoeur's unique contribution to the analysis of narrative emplotment in his theory of a "threefold mimesis" (prefiguration, configuration, and reconfiguration). I have drawn mainly from his first volume.

God is not to be found in nature because that requires faith and faith is a gift of God. Scientists are uneasy because intentionality implies not only design, but a designer. Being sensitive to both concerns, let us consider Leonardo da Vinci's famous painting *The Last Supper.* If an uninformed person sees the painting, the painter cannot be identified. The painting could be appreciated for its beauty and artistic accomplishment, but the artist would remain unknown. Only those who already know da Vinci would be able to recognize the painting as by da Vinci. For the world to be known as creation, so it is argued, one must first know the creator. I agree with the sentiment that a theology of nature (how wonderfully mysterious is the universe) cannot become a natural theology (there must be a divine creator). But da Vinci's *Last Supper* is not just a landscape, it is a painting with characters and their posture, position, eyes, and hands hint at a story. The figure of Jesus elicits further ruminations. No, you cannot know it is the Last Supper, but you do know it is not just any ordinary meal. You cannot know the intention of the painter, but you are enticed to ask further and look elsewhere for other similar paintings in order to learn more about the character of the artist and his subjects.

Characters

In the narrative truth I am proposing, characters are essential to the plot. They are the heroes, the geniuses, and the saints whose lives are intentional and transparent. They have a high degree of self-understanding, purposeful motivation, charisma, and integration. Like David Tracy's understanding of a classic, their lives have a surplus of meaning that spills over. At times, the character is the paradigmatic event, the catalyst for events to happen. Without particular characters, the same narrative truth could not be told. In other times and places the paradigmatic event is the congruity of individual and event: the right person in the right place at the right time. There are other occasions when the event or events sweep the character along. This is how history unfolds.

Novelists even profess that sometimes the plot, the character, or the place "takes on" a life of its own and takes the story in its own direction.

Scientists who write histories of the universe or of the biological world do not include characters with intentions. To do so, they believe, would be to impose upon the world a pattern or purpose that is not verifiable. They would readily admit, however, that they are searching for patterns and finding what is universal in cause and effect relationships. Biographers, on the other hand, limit their scope to events that cohere around an individual. Historians have a broader scope in that they include personalities, historical events, and the landscape within a chronological period. With delightful variety, novelists combine plot, character, and place to entertain, teach, motivate, record, and transfix. What separates the experimenter and the storyteller is not their excitement about how the story will turn out—this they share—but how they will explain and describe causal connections.

Religious faith finds a different pattern. The difference is not a matter of having a unique epistemology in order to see a different world (ontology). Rather, religious faith wants to tell a story that includes human beings who believe that there is a divine intention for the world and for human existence and believe that certain events and human beings disclose that intention. The philosopher Alasdair MacIntyre states the problem this approach presents:

> In the Bible men go on journeys, suffer greatly, marry, have children, die, and so on. So far no difficulty. But they go on their journeys because God calls them, suffer in spite of God's care, receive their brides and their children at the hand of God, and at death pass in a special sense into God's realm. . . . This reference to God introduces all the difficulty. What is said of God is again familiar enough. God calls, God hears, God provides. But these verbs appear to lack the application which is their justification in nonreligious contexts. . . . The name "God" is not used to refer to some-

> one who can be seen or heard, as the name "Abraham" is, and when descriptive verbs are used to state that God's call is heard, it is not ordinary hearing that is meant.[16]

Putting aside such anthropic gestures of God's talking and feeling, each story was preserved because a pattern of divine initiative and response was recognized. This pattern, while identifiable, cannot be separated out except formally, because it coincides with the ordinary mix of cause and effect. The human characters in these stories believed that they were responding to a divine initiative. Abraham did not leave Haran for the land of Canaan just because he thought it was the thing to do; he left believing he was responding to God's call. Significant events in Abraham's long journey served to reveal and confirm a purpose not his own. The plot unfolds, revelation happens, and—at least for some people—God is in the thick of it all. Too much emphasis has been placed on revelation as the paradigmatic event itself, namely its supernatural causation. Particular events do indeed stand out and are memorable, but always and necessarily because they are connected by memory with other events.[17]

Place

At times, we have sorely neglected the importance of the place or setting for narrative truth.[18] Storytellers have not. They seem to understand that the plot and the characters are inexorably intertwined in such a way that narrative truth must include both. "Tell me the landscape in which you live," declares Ortega y Gasset,

---

16. MacIntyre, "The Logical Status of Religious Belief," in *Metaphysical Beliefs*, ed. Toulmin et al. (New York: Schocken Books, 1957), 133, 178–79.

17. In exploring Barth's understanding of revelation, Graham Ward comments: "Revelation (which can only be appropriated retrospectively, as a memory, a 'looking back') enables us to recognize that language is analogical" (Barth, *Derrida and the Language of Theology*, 15). There is both an objective and subjective, an ontological and epistemological dimension to revelation, and Ward is emphasizing the epistemological dimension when he writes that revelation is appropriated retrospectively.

18. Edward S. Casey has not neglected the importance of place. See his *Getting into Place* (Bloomington: Indiana University Press, 1993).

"and I will tell you who you are." Landscape not only has a way of molding personality, it also shapes the plot. The setting of a story is more than the geography. When is a rock a rock, and when is it a cliff? Consider the difference between such classics as *Ulysses* and *Wuthering Heights;* in the former, the geography represents the challenge of getting there and back, while in the latter the moors have left their imprint on the very souls of the characters. Dickens, Faulkner, and Hemingway had an acute sense of how plot, character, and place are inseparable. J. R. R. Tolkien's trilogy *The Lord of the Rings* is another work that is as much about the settings as it is about the actions of the players. In his popular Dune series, Frank Herbert weaves the elements together so that without any one element, the story falls apart. The planet Dune is far more than a stage for the story—it is as much a part of the plot as the human characters who strive against each other. In Ahab's titanic struggle with the great white whale, Ike McCaslin's encounter with the great elusive bear in "The Bear," or the ancient mariner's conflict with the sea, nature is not a place or a stage but an event, a character, a truth that can be told only in narrative form.

We must turn to science fiction writers to appreciate the universe as the larger setting for the forces of good and evil. Science fiction realizes that we who live on the planet Earth are the creatures we are because of where we have evolved, and the kind of place where we have evolved is interconnected with the evolution of the whole universe. A different planet, a different kind of evolution, a different story. Yet, there may well be some commonality across galaxies that will make it possible to speak of a larger narrative truth.[19]

Bacon and Descartes turned the screw in a different direction when they created a mindset that understood the physical world as something to be worked on. The Enlightenment may have replaced an Aristotelian teleological ontology with a historical dialectic as the interpretive framework for understanding who we

19. Arthur C. Clarke and Gentry Lee present us their understanding of that commonality in their story *Rama Revealed* (New York: Bantam, 1994).

are, but the dialectic was fatefully misplaced as man over against nature or simply as man in history (the theological viewpoint). Paul Santmire documents the ambiguous way theologians have treated the natural world. All too often nature is portrayed as simply the context for God's redemption of human beings, and that context has little to do with nature and everything to do with history. In Karl Barth, for example, theology is fixed on the story of the human creature. Nature becomes a backdrop for the drama of the God-human encounter as it takes place in historical time. Nature is made the servant, so to speak, of the history of salvation. In Santmire's insightful interpretation of Teilhard de Chardin's theology, we realize how thoroughly anthropocentric redemption had become. Ironically, although nature and the cosmos played a critical role in Teilhard's understanding of redemption, in essence we find that he has only enlarged the stage since the cosmos is featureless and sterile (*The Travail of Nature,* 155–73).

In a postmodern era such as ours, the story of human existence has a new setting, the unimaginable vastness of the universe. As science unravels the story of the birth of galaxies, we know that we must tell the story of human existence differently than our premodern and modern ancestors. The characters seem to be human in very much the same way, but the plot with its universal themes is immensely more intricate and interwoven. The better we can locate ourselves in the universe as a whole, the more complete the truth we narrate will be.

– SECTION 13 –
DECISIVE EVENTS

This section, along with the next one concerning revelation, proposes the category "decisive event" in order to provide a meeting place that is acceptable to both science and theology. Even before we speak about events that are revelatory, the ground is prepared by finding decisive events in both nature and history. The presence of significant events in history and in nature is fundamental to speech about the narrative truth of the universe.

While the distinction is not crucial to my argument, there is a difference between significant events and decisive events. A significant event is any event that introduces a new configuration of meaning. Not every significant event becomes decisive for the fate of one's life or for human life or for the course of evolution. We know, for instance, that the evolutionary path of the hominids included one or more dead ends, such as *Australopithecus africanus*. New species and new particles come into existence and pass out of existence without the same decisive impact as other species and particles. However, it was decisive that nucleic acid molecules became organized in such a way that they could transmit biological information. It was decisive that plants acquired an ability to convert radiant solar energy into chemical energy. Likewise, it was a decisive event when helium cores collapsed and fused into carbon, a stable state necessary for stars and other heavenly bodies to form. Such decisive events are combined into what is known as the anthropic principle, because they are the known critical points that were absolutely necessary for the evolution of human beings on this planet.[20]

What Is a Decisive Event?

It would be glib to say science knows perfectly well what an event is, while theology is burdened with the language of the miraculous and supernatural, which serves to place those events in a category of their own. Every event, whatever its significance, is isolated at the expense of denying that it is continuous with its past and its future. Methodologically, modern science is predicated on the belief that the universe can be understood by isolating larger events into smaller and smaller fundamental events and that each event

20. See J. D. Barrow and F. J. Tipler, *The Anthropic Cosmological Principle* (Oxford: Clarendon, 1986) and John Gribben and Martin Rees, *Cosmic Coincidences* (London: Heinemann, 1989).

is determined by a set of causal laws. Moving in the opposite direction, much of science in the last half of the twentieth century has been looking for the forces that bind together all events. The standard experiment is nothing less than a cut made into the universe in order to isolate a cause and event sequence. Along with this, arguments for new and subtler interpretations of nature arise from the larger integration of the connectedness of the universe. Jacob Bronowski writes: "We make a cut. We put the experiment, if you like into a box. Now the moment we do that, we do violence to the connections in the world" (*The Origins of Knowledge and Imagination,* 58–59). Consequently, the wider integration of connecting systems is ignored. These systems of interdependency cannot be omitted, however, because science cannot function without large-scale concepts such as gravitation, mass, energy, enzymes, genes, minds. There would be no disagreement today that science has just as much to learn by understanding cause and effect at ever-larger levels than by isolating the minutest cause-and-effect sequence. The larger the event, such as the beginning of a new species or the assassination of President Kennedy or the discovery of DNA, the more it is necessary to understand that event in light of its past and future contexts.

At first, it seems that scientific theories present a formidable wall against narrative truth because natural selection, chaos, and chance rule out any connected meaning. Physicists, however, must begin with a puzzle.

> The universe is made up of units which are highly active, call them atoms or what you will. The activities of these units are highly independent, one of the other. There is nothing in the nature of these units to keep them from frustrating and destroying one another and producing a hopeless confusion. Yet as a matter of fact they do not fall into such confusion. Doubtless there is plenty of confusion and frustration, but at the same time there is a very high degree of order, of mutual adjustment and mutual support between all these seemingly

> diverse and relatively independent activities. How can we account for this mutual support and mutual adjustment? Whence does it come? It does not come from the atomic units. It cannot come from the higher organization of these units, such as animals and men and societies, for these latter should never have arisen if the microscopic units were not adjusted and organized sufficiently for those more complex bodies to arise. Whence then comes this mutual adjustment and order?[21]

Alfred North Whitehead was one of the first scientist-philosophers to be particularly interested in the problem of indeterminacy and order, unity and multiplicity, and their theological implications. Creativity is what brings actualities into being. Chance and necessity, as Arthur Peacocke suggests, are the means by which creative potentialities are being "run through" or "explored." Natural selection sifts through these chance occurrences and, by using principles not entirely understood, allows some to die and others to exist by maintaining their identity and/or by transmitting their DNA. Extinction is the price paid by evolution for groping. Whitehead and Teilhard de Chardin were right in not allowing science to insist that chance is unconstrained. One step, then, toward a paradigm of decisive events is to recognize that chance is constrained by order, that it accumulates and leads toward certain storylines. Stuart Kauffman has made it his profession to explore the laws of self-organization and complexity that are a "natural expression of a universe that is not in equilibrium, where instead of the featureless homogeneity of a vessel of gas molecules, there are differences, potentials, that drive the formation of complexity" (*At Home in the Universe: The Search for Laws of Self-Organization and Complexity*).

Holmes Rolston writes: "Some events may not be intelligible or interesting or significant, some are like mutations and worth-

21. Nelson Wieman quoted from Charles Hartshorne and Creighton Peden, *Whitehead's View of Reality* (New York: Pilgrim, 1981), 87.

less trial ideas that, though parts of groping stories, get selected against. In the background of the storylike signals there is meaningless noise, events without coherence or point, just as there are noise and randomness in the background of theories and causal laws" (*Science and Religion,* 277). Plot, then, does for history what theory does for evolution. "One has to formulate plots to detect the stories, no less than one has to construct theories to find causes systematically" (278). It is difficult to argue with Rolston's conclusion that one does not have a complete explanation until one has the stories, although minor "subroutines of explanation may not require story" (277–78).

Modern science has been deeply reluctant to say it configures time as narrative, because it is hesitant to conclude there might be direction or purpose in the sequence of evolving events. But that reluctance is slowly breaking down as science matures. An important part of this maturity is the acknowledgement that laws are inadequate if we are to understand a totally connected universe. To understand such a universe requires a longer, broader sweep of events. Field theories and system theories, for example, began the broadening process. Whitehead gave a philosophical foundation for understanding physical occurrences as connected events.[22] For Whitehead, "concrescence" is the process in which a universe of many things acquires an individual unity. He concluded: "From this point of view, he [God] is the principle of concretion—the principle whereby there is initiated a definite outcome from a situation otherwise riddled with ambiguity" (*Process and Reality,* 523).

Disciplines such as cosmology, meteorology, paleontology, and oceanography have been revived by the rediscovery of time. Evolutionary biology, for instance, is a relatively new discipline that considers that an organism is understood fully only when it includes a history of the changes of its genetic program through time

22. Alfred North Whitehead, *Process and Reality* (New York: Macmillan, 1957), 101–8, 321–22; Arthur Peacocke, *Creation and the World of Science* (Oxford: Clarendon, 1979), 70.

and the reasons for these changes. Physics, likewise, has moved steadily away from studying particles as self-contained objects. At the level of the microworld, the particle is always an event that is part of an ever-widening circle of interrelated events. At the other extreme of size, the macrocosm of stars and galaxies, the approach has become to write an evolutionary history of the birth of the universe. Again, we find that descriptive laws of the physical universe must also satisfy the requirements of evolutionary history. Without even interjecting questions about direction and purpose, each organism and each star has a story of its own. To know the history of one star we must know the history of its galaxy, and so forth. To the extent that we are able to understand the larger story, the better able we are to assess if the human story is a more focused expression of these more general processes.

The more we enlarge the story, the more natural narrative truth becomes. Is it then plausible to speak of an ontology that invites narration? Rolston writes, "Sometimes a thing needs to be understood not merely immanently, in terms of what it now is in its own-being, but in terms of what it is becoming, as a link in a story" (*Science and Religion*, 300). Rolston uses the word "supernatural" in the sense that some emergent steps transcend the natural and "outdo our capacities for analysis" (301). While this is a creative way to integrate the word supernatural, it is not as helpful as saying some events transcend or "stick out" from what is ordinary and forgettable. We are not trying to identify specific forces that transcend the physical, the biological, or the sociocultural. What we see is the effect of evolutionary processes that are more than blind groping. Something acts to coalesce and compound the fortuitous and "macrohistory draws the microhistories after it, although the microhistories (genetic mutations, individual careers and choices) emit the novelties that perfuse the macrohistory" (Rolston, *Science and Religion,* 302).

The universe, then, is a network of events where not all events have the same significance. There are those that stand out. A decisive event, whether historical or evolutionary, is the juncture where the flow of the ordinary converges to create something radically new. Decisive events are determined by how much of the past they carry into the future, how significantly they alter the flow of the ordinary, and how unique they are in and of themselves. The biologist will speak of a unique event, the physicist of a singular event, the historian of a paradigmatic event, the psychologist of a turning point, the theologian of a conversion. Despite the difference in language, the phenomena they describe are all memorable; they are not lost in the ordinary flow of events. There are many different reasons why they may be memorable: their singularity, their size or force or visibility, their significance, their depth, or their novelty.[23]

Jacques Monod said the scientist is not obligated to explain why a unique event has occurred, but as Ernst Mayr counters, "uniqueness, indeed, is the outstanding characteristic of any event in evolutionary history" (*The Growth of Biological Thought,* 71). In other words, significant events cannot be ignored just because they cannot be explained. The universe is orderly and our theories may be aesthetically elegant, but order and elegance tell us little. For science and theology to tell their stories, they must look to the unfolding of the unique events that push beyond continuity to narrative truth.

Characteristics of Decisive Events

Without decisive events, there would be no drama, no adventure. In the decisive and unique event, the ordinary becomes interesting, chance is channeled into direction, and the truly new is created. While it is more accurate to say that reality is constituted by a continuity of becoming, we know that we experience the world not as a seamless continuum but as the becoming of

---

23. Examples include Big Bang, 4 fundamental forces, particles; nebulae, planetary systems, galaxies; liquids, nucleotides, DNA/RNA, proteins; cells, plants (photosynthesis), vertebrates, *Homo sapiens;* Buddha under the tree, Jesus of Nazareth, Albert Einstein; Mycenaen civilization, use of iron for tools and weapons, first atomic bomb, the Holocaust.

distinct unit-happenings. If we were to experience life as a continuum, then eating breakfast would not be an event but an infinite number of indistinguishable experiences. Whitehead and Ira Progoff have called our attention to how and why life moves forward by new unit-happenings.[24] Whitehead and Charles Hartshorne observe that everything new proceeds out of a creative synthesis whereby the many flow into a new one producing a new many, and so on forever. While everything has the potential for all future actualities, not everything in fact contributes to what becomes, and even fewer potentialities contribute to a decisive event. It would appear, nevertheless, that nature reaches new levels of complexity only by way of decisive events. To use the language of Whitehead, in the creative act "the many become one and are increased by one," but in the decisive event the many become one and the direction of becoming is changed in a memorable way. If the decisive event cannot be denied, the way is opened to converse about providence, purpose, revelation, and God.

Following the lead of H. Richard Niebuhr and Whitehead, Van Harvey describes a decisive event as paradigmatic because "it so captures the imagination of a community that it alters that community's way of looking at the totality of its experience" (*The Historian and the Believer,* 253). Such events call forth a new understanding and a resolve. The assassination of President Kennedy, rather than his natural death, became such an event because it created Kennedy as a symbol of a new style of patriotism and leadership. His murder intensified a moment in time while giving it universal significance (*The Historian,* 257). A paradigmatic event is one that freezes reality at a moment of time but allows memory to continue to thaw it with ever-wider circles of universal significance. Such events become legends, myths, and symbols serving to reorient our understanding of reality.

---

24. From the very different perspective of journal writing, Ira Progoff defines an event as "the particular occurrences that take place with meaningful impact on the movement of our lives. Events of this kind call attention to themselves by their dramatic quality, by the questions they raise about the mysterious movement of human existence." See Progoff, *At a Journal Workshop* (New York: Dialogue House Library, 1975), 212.

Another way to understand a decisive event is to use David Tracy's analysis of what constitutes a classic. "My thesis is that what we mean in naming certain texts, events, images, rituals, symbols and persons 'classics' is that here we recognize nothing less than the disclosure of reality we cannot but name truth" (*The Analogical Imagination,* 108–9). What distinguishes a classic or a decisive event, then, is its surplus of meaning. A popular song is not likely to endure for centuries because its meaning is quickly mined. It may have universal appeal because it strikes a ubiquitous human chord, as country music does, but it lacks sufficient complexity to draw us back to it repeatedly, each time discovering and reappropriating something new. A classic, therefore, is characterized by its permanence and its demand for a plurality of interpretations.

John F. Haught adds yet another perspective. In his undervalued book *The Cosmic Adventure* (1984), he argues for reinstating teleology—not as the cosmos aimed at consciousness, but toward beauty. Haught prefers this reading of evolution to the consciousness-oriented theologies, such as that of Teilhard de Chardin, because it is more comprehensive. He also prefers an aesthetic reading of evolution to an ethical type of teleology, because the latter argument gets entangled in justifying a good creator over against the capricious, cruel face of nature. Beauty exists because of the tension between order and vulnerability, between permanence and extinction. Beauty is threatened on two sides: by chaos on one side and by monotony or triviality on the other. The aesthetic, Haught writes, is "a balancing act between the extremes of chaos and banality. It is precarious, and therefore is precious" (102). Such is the beauty of a real rose over an artificial one. The beautiful person is also one who stands out, whether we call him or her a hero, a genius, or a saint.

> The hero or genius arouses our admiration for essentially the same reason as a work of art. Heroism, for example, is beautiful because it is the result of integrating a multiplic-

> ity of contrasting experiences (strength and frustration, joy and tragedy, rebellion and resignation, life and death) into a unity of a single person's story. Genius is beautiful because it requires the integration into a creative unity of a multiplicity of ideas, feelings and experiences that could lead to madness in a narrower personality. (*The Cosmic Adventure,* 105)

Along with beauty, the universe exhibits the quality of adventure. Because the universe is forever experimenting with fresh forms, it makes life interesting and adventurous. And God's place in all of this is to lure the cosmos toward adventure, "constantly awakening it from the inertia that would fix it into any given order."[25]

History and evolution, then, cast up events that are disclosive because they both intrude and protrude with a surplus of meaning. They are revelatory because they raise such profound questions that reference to God seems necessary.

Order, Design, Story

There have been many attempts to order the events of evolution into a pyramid-like structure of decisive events. For example Scott Peck, in his immensely popular book *The Road Less Traveled,* summarizes the flow of evolution as running against entropy. At the apex is man, the most complex being; viruses, the most numerous but least complex organisms, are at the base (265). Mostly these efforts have been presumptions, because any arrow of direction is inferred and our prejudices quickly intrude.[26] Without the presumption of a pyramid, a list of decisive events merely calls attention to the fact that neither evolution nor human history flows

25. John Haught, *The Cosmic Adventure,* 133. Haught returns to this theme ("A God for Evolution") in his *God After Darwin: A Theology of Evolution,* chap. 6. Darwin's theory of evolution is no longer a threat or a "dangerous idea" (Daniel Dennett) but an understanding of the world that should be taken into the very center of theology's reflection on the meaning of life, God, and the universe (33).

26. Compare Daniel C. Dennett's discussion of "Darwin's Assault on the Cosmic Pyra-

uniformly like a lowland river. The course is uneven and more like a mountain stream with periods of rapid flow, surges, and diminutions as well as changes in direction due to fallen trees and shifts in the ground. The terrain, the information that is passed on, the "upward" influence of the organizational hierarchy in which it is embedded, the intervention of "human hands," and so forth determines where evolution goes and how it flows.

Yet any ordering cannot ignore what Teilhard de Chardin referred to as a "privileged axis" or a "convergence of complexity," or what Theodosius Dobzhansky calls "evolutionary transcendence."[27] Evolutionary transcendence refers to those moments or periods when a radically new unity emerges or a new path is "canalized." C. H. Waddington introduced the concept to describe the development of an organism leading to a preferred pathway that he named the *chreode*.[28] Genetic changes and environmental perturbations, coupled with the opportune moment, can push the course of development above the threshold into another chreode. Rupert Sheldrake's diagram (Fig. 13.1 on p. 236) captures both the process of canalization and the presence of the chreode.

The image that comes to mind (Fig. 13.2) is the familiar sight of water drops running down a large, slightly inclined plane of glass. Some of the drops diminish and quickly disappear (evaporate),

---

mid" and "The Principle of the Accumulation of Design" in *Darwin's Dangerous Idea*, 64–73. Dennett's own pre-Darwinian pyramid looks like this.

God
Mind
Design
Order
Chaos
Nothing

Dennett's quarrel is not that the universe exhibits chaos, order, or mind but that purposeful design emerges from order and time ("Give me order and time and I will show you design" was Darwin's mantra). In principle, then, would Dennett rule out a hierarchy of order?

27. For a discussion of Teilhard de Chardin and his own thoughts about evolution as directional but not necessarily directed, see Theodosius Dobzhansky, "Teilhard de Chardin and the Orientation of Evolution: A Critical Essay," in Ewert H. Cousins, ed., *Process Theology* (New York: Newman, 1971), 229–48,

28. Rupert Sheldrake, *A New Science of Life* (Los Angeles: J. P. Tarcher, 1981), 50–51.

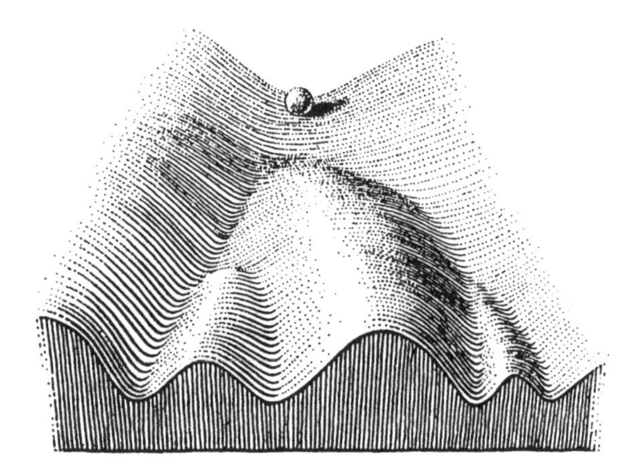

Figure 13.1. An epigenetic landscape illustrating the concept of the chreode as a canalized pathway of change.[29]

but other drops merge and gain momentum and visibility and that momentum may result in further mergers.

For the moment, let us say only that this is the stuff of narrative truth. Concepts like chreode, direction, purpose, destiny, and emplotment are ways to talk about time as it coagulates. Time itself is meaningless and neutral, like the sand running through an hourglass. Time is memorable only when it is configured and there are many ways to configure it, chief among them are theological and scientific ways. In order for a story to be told, for a history to be written, there must be events that call attention to themselves by their dramatic quality, by the questions they raise about human existence, and by the meaning that flows from them.

The creative, adventurous dimension of the universe takes place when the simple becomes the complex, the inanimate becomes the animate, consciousness becomes self-conscious, and "Homo" reaches to touch Mars. From time to time and from place to place, chronological time is punctuated with *kairos*—that oppor-

29. Diagram from R. Sheldrake, *A New Science of Life* (Los Angeles: J. P. Tarcher, 1981), 51, which in turn was taken from C. H. Waddington, *The Strategy of the Genes*, chap. 2.

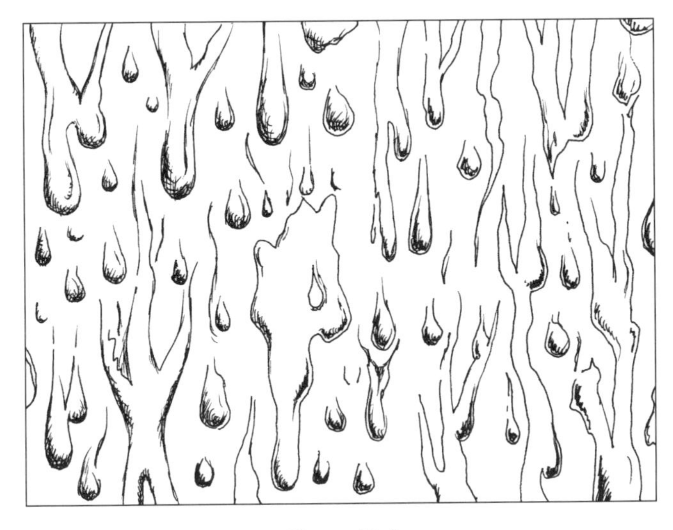

Figure 13.2

tune moment when a process "climbs above" the continuity of chronological time. Time is forever pregnant with opportune moments waiting for the decisive to happen. For the most part these moments never happen and pass into the oblivion of continuity. Whether by small or momentous acts of transcendence, a course is generated that looks like direction and suggests a story. And in all of this there is the struggle between life and death, good and evil, creation and destruction, the spiritual and the material. It is this struggle that Charles Darwin intuited as the most troublesome aspect of his theory of evolution.[30] Others rec-

30. Adrian Desmond and James Moore title their remarkably detailed and holistic account of Darwin *The Life of a Tormented Evolutionist* (New York: W. W. Norton, 1991). It was not just that Darwin could see little evidence of progress, but that he experienced, in both his personal life and in nature, the same wasteful, capricious course of events. The source of Darwin's torment was that he could not reconcile the traditional beliefs of the Church of England and his respect for his wife as a true believer with a scientific theory to which he had publicly committed himself and defended with all of his intellectual strength. The death of his beloved eldest daughter Annie at the tender age of nine was his first "bitter and cruel loss." Desmond and Moore comment: "This was the end of the road, the crucifixion of his hopes. He could not believe the way Emma believed—nor what she believed. There was no straw to clutch, no promised resurrection. Christian faith was futile" (384). Darwin never considered the possibility of "a theology of the Cross" and therefore never thought to look at the decisive events in his life, including those that were tragic as well as beneficiary, in a different light. The reason, of course, was the preemi-

ognized the tension and the paradoxes, but backed away from looking them straight in the eye. Even today, it would seem, only science fiction writers give full reign to the storylike underpinnings of all that has happened and could happen on this cosmic journey.

nence of a theology of special creation and divine providence. Whatever theology he had learned and whatever Bible Darwin read, it did not include suffering and crucifixion at the very core of the Christian faith. See Haught, *God after Darwin,* who rightly critiques the "one-sided appeals to the idea of God as an 'intelligent designer' " and thereby renders "the issue of theodicy [how to justify God's existence as "the lover of our souls" with the fact of suffering and evil] all the more intractable" (45).

Chapter 6

The Truth We Seek

On the very brink of human possibility there has, moreover, appeared a final human capacity—the capacity of knowing God to be unknowable and wholly Other; of knowing man to be a creature contrasted with the Creator, and, above all, of offering to the unknown God gestures of adoration. This possibility of religion sets every other human capacity also under the bright and fatal light of impossibility.

—KARL BARTH, *The Epistle to the Romans*,
translated from the 6th edition by Edwyn C. Hoskyns, 250

I believe that faith in Ultimate Reality can make all the difference for a life of resistance, hope, and action.

—DAVID TRACY, *Plurality and Ambiguity*, 110

– SECTION 14 –
REVELATION RECONSIDERED

Introduction

A fair summary of the previous chapter would be that God's activity is to be located in those significant and decisive events of history and evolution and in the pattern they weave. However, to call this pattern of events revelatory in a postmodern context, it is necessary to discard much and to redefine how the word "revelation" is used.[1] To be acceptable and plausible, a narrative story of the universe must incorporate the theological assertion of revelation and at the same time not violate the scientific bar against ad hoc presumptions. In the previous section I set forth the concept of "decisive event" as a basic category in the conversation between science and theology. Since the reign of empiricism began, theologians have struggled to define revelation so that it might regain a respectable place in the academic community. As a theological category, it has been undermined on two fronts. As the primary or reigning mode of knowing, empiricism made it difficult for theology to rely upon revelation as a unique kind of knowing. If "to have faith" or "to be inspired" meant a method of knowing that did not include verification and generate consensus, then scientists wanted nothing to do with it. On another front, science was asserting a mechanistic world that operated according to universal law and a biological world that evolved by chance; natural selection simply did not have a place for "acts of God," whether natural and in harmony with universal law or supernatural and against the grain of natural law. An entirely consistent and self-sufficient universe forced theology to rethink

1. Ronald F. Thiemann begins his book about revelation by reviewing the situation this way: "Indeed, most discussions of revelation have created complex conceptual and epistemological tangles that are difficult to understand and nearly impossible to unravel. A sense of revelation-weariness has settled over the discipline and most theologians have happily moved to other topics of inquiry." See Thiemann, *Revelation and Theology*, 1. For a different assessment, see Gabriel Fackre, "The Revival of Systematic Theology," *Interpretation* (July 1995): 229–41.

two cornerstones in its understanding of revelation: providence as the sign of the sustaining and guiding hand of God, and miracles as the evidence of the power of God to intervene in his own creation.

It must be remembered that what revelation did, besides positing certain supernatural events, was to provide the widest cosmological framework within which the story of human life and the world was told. By framing the human story with a beginning and an end, revelation gave human beings confidence in the providential course of natural and historical events. However, science was gradually providing a different account of the universe. First came the universal mechanism; second, a purely historical account of the origins of the earth and its creatures. The new sciences (geology, paleontology, and biology) not only displaced the Bible as a source of revealed knowledge, but also raised questions such as: How could God be a cause? Langdon Gilkey summarizes: "The divine activity, and hence religious statements about it, was, so to speak, banished from the contemporary universe and moved safely out of the range of matters of fact then available to scientific inquiry" (*Religion and the Scientific Future,* 8).

Thus, revelation was discredited both epistemologically and ontologically. How was theology going to "keep up" with science when the events it championed were known by a method of faith and inspiration? Karl Barth forced the issue and theology split between choosing to speak of God only as the subject who is never to be objectified or of God who is objectively revealed in the historical Jesus of Nazareth and in the events of salvation history.[2]

Even when you take into account everything that has changed in a postmodern situation, theology must nevertheless tackle the same two difficulties; namely, how to speak about providence and

---

2. These two were not the only positions, but they have been dominant in Protestant theology since 1930. See Gareth Jones, *Critical Theology: Questions of Truth and Method* (New York: Paragon House, 1995), chap. 1, for his discussion of the divide between Harnack and Barth, which was no less than a question about how scientific theology was going to be.

miracles, or in a more preferable language, how to speak about God's self-revelation within the flow of chronological time and the interruption of time in the *kairos*, or providential moment.

We are at once on firmer ground if we keep before us the Christian belief in a triune God—God the creator, God the redeemer, and God the sustainer. This is a necessary first step because it compels our language about revelation to be multifaceted. The Church, defending a monochromatic doctrine of revelation, was led to excess, reductionism, and one-sidedness by emphasizing one dimension of God at the expense of the others. Ted Peters asks whether Christians have become monotheists (*God as Trinity*, 37–42). Pietism has consistently over-emphasized the individualism of Christ the redeemer (a fall-redemption form of spirituality), highlighting revelation as a hot-flash experience. Biblical theology of the early twentieth century became preoccupied with the historical "acts" of God as a historical inbreaking of Yahweh. Liberalism of the nineteenth and twentieth centuries, including contemporary liberation theology, virtually forgets the Christian understanding of God as three identities with one substance. It would certainly appear that scientists think Christians worship a God who is an autonomous, solitary, self-contained, apathetic, absolute monarch. One thinks of Einstein who could not abide such a God and Tillich who moved heaven and earth to turn Christian theology away from a monotheistic ontology.[3]

Revelation in Context

Just as there can be no bare scientific fact, so there can be no bare revelatory "act." An experience of the divine may be singular, but it is nevertheless an experience with important preparatory antecedents and important posterior reflections. The supposed

3. Roy D. Morrison understands Tillich as wanting to reorient theology's understanding of both ontology and epistemology away from the classical worldview that required objective or supernatural referents, of which God as a personal being was paramount. See Morrison, *Science, Theology and the Transcendental Horizon: Einstein, Kant and Tillich* (Atlanta: Scholars Press, 1994), 129–32.

singularity of a historical event is always a happening artificially removed from its context, whether it is a paradigmatic event such as the exodus or the resurrection of Christ, or a personal conversion such as Paul's encounter with Jesus Christ as the Messiah. As a singular, isolated event, revelation is nothing more than a secular, ordinary event. The crossing of the Red Sea, for example, did not have redemptive significance for the pursuing Egyptians, because they did not share the Hebrew's understanding of salvation history. For the Egyptians it was just one defeat among other defeats that would quickly be forgotten (not to be recorded). For the Hebrews, however, their deliverance had a deep and pervasive meaning, because it was part of an overall pattern of events (providence) beginning with the anonymity of being "no people" and moving toward becoming a nation settled in the "promised land." The actual recording of their deliverance at the Red Sea was a progressive event that became even more significant as the author (speaking for the community) reflected back on the entire series of events that by the post-exilic period (sixth century) had become a pattern of redemption. If time is to be remembered, it must be reconfigured. The Bible is a theological configuration of time. There is also a scientific way to reconfigure time; astronomers do it consistently. In order to tell a story, the plot lines are traced backward to establish an intelligible pattern of relationships and forward to project an intelligible pattern of what to expect next.

In *Blessing in the Bible and the Life of the Church* (1978), Claus Westermann joined a growing number of scholars who criticize the hegemony of salvation as deliverance over salvation as blessing. As an Old Testament scholar, Westermann argues that the Bible gives witness to a God who is not only a deliverer but also a dispenser of blessing. The contemporary church, however, has nearly ignored the latter. He writes:

> From the beginning to the end of the biblical story, God's two ways of dealing with mankind—deliverance and bless-

> ing—are found together. They cannot be reduced to a single concept because, for one reason, they are experienced differently. Deliverance is experienced in events that represent God's intervention. Blessing is a continuing activity of God that is either present or not present. It cannot be experienced in an event any more than can growth or motivation or a decline of strength. (*Blessing*, 3–4)

Westermann's critique of Western Christianity is an accusation that a kind of reductionism has taken place. The saving activity of God should not be spiritualized into an encounter between God and the human soul. Neither should it be solely temporalized and dramatized as "mighty acts." God's self-revelation is both the momentary experience of forgiveness and the coincidence of many historical incidents that came to be known as the exodus or the crucifixion. In Scripture there is both the language of deliverance and the language of blessing; one is definitely historical and the other is usually regarded as "unhistorical." There is the God who saves by coming and there is the God who is present by bestowing and nurturing. The priest becomes the primary mediator of God's continuous blessing, while the prophet and the charismatic leader mediate God's dramatic inbreaking. The complete story of Israel and the full story of Jesus Christ are constituted by both kinds of saving activities. Westermann, then, reminds us that the Trinity is the necessary symbol that holds together in tension the redemptive nature of God the Son and the sustaining nature of God the Holy Spirit.

Translating Supernaturalism

In the Christian tradition, God's interaction with the world is understood as supernatural and miraculous.[4] Here my argument will be against those translations that submerge the divine activity

4. For another attempt at translating supernaturalism, see Rolston, *Science and Religion*, 298–306.

so that it disappears into the flow of history, and contrary to those who think we must identify divine activity as causal. It is no understatement that supernaturalism has become a thorn in the flesh of theology. Scientists often see it as one of the trademarks that makes theological inquiry suspect. Scientists expect Christians to defend revelation as the singular and the exceptional. In the face of considerable hostility, Charles Darwin remained steadfast in his refusal to admit acts of special creation. In a letter to the geologist Charles Lyell shortly after the publication of *Origin,* he wrote: " 'I would give absolutely nothing for the theory of Natural Selection, if it requires miraculous additions at any one stage of descent. . . . If I were convinced that I required such additions to the theory of natural selection, I would reject it as rubbish' " (quoted from Dennett, *Darwin's Dangerous Idea,* 60). Darwin's understanding of natural selection was analogous to a universal sieve that allowed only the smallest of chance occurrences to pass through. Consequently, there was no place for decisive events, no matter how they came about.

Peter C. Hodgson's *God in History* represents a healthy attempt to find a middle ground between interpreting God's presence in history as a generalized influence or as the individual personal acts of intervention by God. He writes:

> My thesis is that God is efficaciously present in the world, not as an individual agent performing observable acts, nor as a uniform inspiration or lure, not as an abstract ideal, nor in the metaphorical role of companion or friends. Rather, God is present in specific shapes or patterns of praxis that have a configuring, transformative power within historical process, moving the process in a determinate direction, that of the creative unification of multiplicities of elements into new wholes, into creative syntheses that build human solidarity, enhance freedom, break systemic oppression, heal the injured and broken, and care for the natural. A shape or gestalt is not as impersonal and generalized as an influence

> or a presence, since it connotes something dynamic, specific, and structuring, but it avoids potentially misleading personifications of God's action. What God "does" in history is not simply to "be there" as God, or to "call us forward," or to assume a personal "role," but to "shape"—to shape a multifaceted transfigurative praxis. (*God in History,* 205)

In one sense, the commonly held notion is valid. For something to be called "revelation," it has to stick out, because that is what makes it memorable and worthy to be taken up again and again. Only certain events are memorable and resist being forgotten. This is their ontological property and their transcendent quality in the etymological sense of the word: *trans* = over + *scandere,* "to climb above the ordinary." They are significant in themselves because they stick out but also because they have the epistemic quality of galvanizing additional significance as they are known. That is, they invite epistemological attention. Other historical events are drawn to them like steel filings to a magnet. Being so constituted, they have the power to inspire and illumine, to move or energize the plot.

In *The Analogical Imagination* David Tracy defines the quality of the classics in every culture: their memory haunts us, they endure, their effects in our lives await ever new appropriations, and they spill over with a surplus of meaning. Tracy, in conjunction with Hans-George Gadamer, points out another quality of the revelatory event. When we participate in a revelatory event, our self-awareness and self-centeredness is lost in what Gadamer calls "a realized experience of an event of truth." "We find ourselves," Tracy writes, in the authentic experience of a classic text or art " 'caught up' in its world, we are shocked, surprised, challenged by its startling beauty *and* its recognizable truth, its instinct for the essential" (*The Analogical Imagination,* 108–13).

The religious classic, nevertheless, has an epistemological-ontological quality of its own, but this is a matter of degree. Revelation is experienced as "an eruption of a power" that is not our own. Our coming-to-know in the religious experience is still a "being caught up," but now it is in and by the power of the Whole. In other words, the "Whole," which is Tracy's way of speaking of God, transforms both the original event and the retrieval of that event, because the power or inspiration behind that event is still present. Epistemologically, the interpreter enters an intensification process that has the ability to liberate; that is, to transform the knowing process so that it reconfigures one's life or the life of the community. The experiences of mystery, awe, beauty, truth, and fundamental trust work their way through our hearts, minds, imaginations, and through them to our wills. On the ontological side, behind the event-like manifestation that is known as a religious classic or as a paradigmatic historical event, is the power of the Whole. Tracy is a little vague at this point, but his sense is clear enough. God is that Whole because only God is the creator of all that is; in particular "acts" of disclosure there is a more intense manifestation "of the whole from the reality of that whole."

Examining one particular story, the biblical account of creation (Genesis 1–3), we must remember that it was not told to explain events but to locate human experience in a framework of larger significance. We could say along with James M. Robinson that the early Christian belief in Jesus as cosmo-creator meant essentially that to be a Christian meant to be "in step with the universe."[5] This is not to say that biblical storytellers were entirely disinterested in causes, but their storytelling was not motivated by scientific inquiry into the minutest cause-effect relationships. To the extent that ancient storytellers included the divine (God) as a causal factor, it was to mark off certain events with special significance, a significance that transcended the mundane affairs of human life. Their urge to know was not to reason inductively or deductively or to unmask nature with analytical, objective analysis. The overriding questions they were asking had to do

5. James Robinson, "The Biblical View of the World," *Encounter* 20 (1959): 470–83, especially 482.

with knowing who we are and where we live—those perennial questions of meaning and orientation. Their abiding interest in what happened and how it turned out persisted through the various forms of storytelling—folktales, legends, allegories, parables, comedy, and tragedy.[6] Thus, even the most gratuitous fable was to be honored because it told them something about themselves.

Finding God among the Causes

There is a sense among Christians that the only way to honor God as creator and redeemer is to assert that God acts quickly and directly.[7] Yet, we know from personal experience that the significance of an event is not immediately available. We cannot know fully what it means or what it will mean. Instead of trying to understand revelation along the lines of dramatic interventions deserving a supernatural cause, I ask that we look at the unfolding of the universe and the unfolding of one's personal life story in terms of a storyline. Events are not significant or revelatory because of some identifiable cause. Rather, they are significant and revelatory because they focus a pattern of events into an identifiable disclosure. Language that relies upon the miraculous and on the supernatural draws attention to the wrong place.[8] Would it not be better to say that some events should be called miraculous simply because they make their appearance against all odds?

Science has a history of expecting theologians to defend revelation as a supernatural intervention by God, and theologians have

---

6. Frederick Buechner's wonderful little book is more than illustrative of the different forms telling the gospel takes. Buechner, *Telling the Truth: The Gospel as Tragedy, Comedy and Fairy Tale* (New York: Harper & Row, 1997).

7. See Richard J. Coleman, *Issues of Theological Conflict* (Grand Rapids, Mich.: Eerdmans, 1972), especially chap. 3 ("The Nature of Revelation: Absolutes vs. Relatives").

8. The language of supernaturalism is thoroughly entrenched in the Bible. What is not so obvious or clear, in spite of many claims to the contrary, is the motivation of the authors. Beyond saying that they are highlighting an event, marking it off as divine or the "hand of God" in stories where God is the principal actor, I am rather uncertain what they thought about its causal explanation. We know that it was prescientific. We do not know when they expected a literal reading or a more metaphorical one, but always a theological meaning was present.

a history of baiting that hook. Scientists have complained for centuries that miraculous explanations explain nothing at all. For Charles Darwin, revelation and the gradualness of the imperceptible were an oxymoron. It is not surprising that some scientists are still pinning this rap on Christian theologians, because there is a long history identifying revelation with the miraculous when the miraculous is an unexplainable cause or a cause that could only be attributed to God. Let me illustrate this by examining Richard Dawkins's *The Blind Watchmaker* (1986) and Daniel C. Dennett's *Darwin's Dangerous Idea* (1995). Dawkins knows a decisive event in only one manner: "fantastic coincidences," and "large chance events" that are "the once in a lifetime/eon role of the dice." But it happens because almost anything can happen given sufficient time. Dawkins finds it more satisfying to postulate a chance event like a gang of 10^{46} monkeys, each with its own typewriter, so that one "would almost certainly" type, "I think therefore I am." Once allowing for the improbable, Dawkins lets us know what matters: "The more we can get away from miracles, major improbabilities, fantastic coincidences, large chance events, and the more thoroughly we can break large chance events up into a cumulative series of small chance events, the more satisfying to rational minds our explanations will be" (*Blind Watchmaker,* 141).

Daniel Dennett leaves no thinker alone if they hint they are resorting to ideal causes, strokes of special creation, mystery, Lamarckism, nonadaptive jumps, skyhook explanations, or you-couldn't-get-here-from-there organs/organisms (such as the eye). Only better adaptations, occasional wedges, and internal cranes move evolution along. Of course the most offensive skyhook is divine intervention that pulls evolution up to a different level.[9] Dennett is correct that skyhook explanations are not explana-

9. As far as I know, neither Dawkins nor Dennett is familiar with the language of process theologians who refer to God as the lure that pulls and pushes evolution to a new level. It would have provoked some new arguments because, while Dawkins and Dennett are thorough and picturesque, they are not especially original.

tions at all, because they end any further discussion concerning the meaning of events.

Dawkins and Dennett cannot exactly sidestep the presence of significant events, although they seemingly do so by arguing about causes. If we disengage ourselves from the fruitless debate that tries to identify some causes as divine causes, do we then argue about the size or the significance or the frequency of decisive events? Dennett offers two illustrations about such a disagreement.

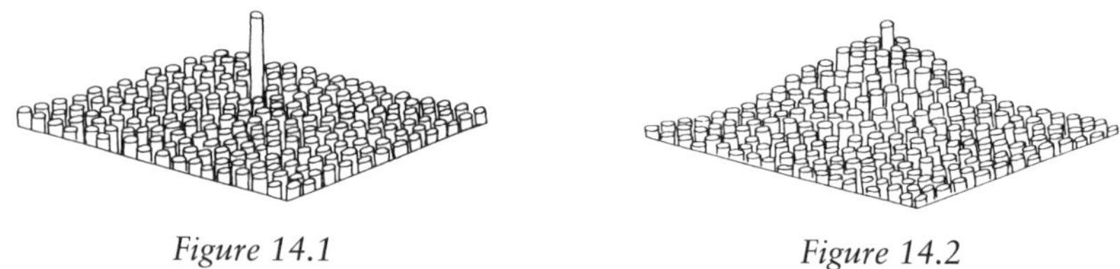

Figure 14.1 *Figure 14.2*

(*Dangerous Idea,* 78, 79)

The altitude or vertical height represents fitness or what Dennett calls a good trick or a better adaptation or better design. I would say it represents the degree that an event is memorable or sticks out. Dennett's argument is that a natural account of evolution only permits a gradual accumulation of better design for the sake of survival of the genes. First, revelation can be about vertical height without resorting to skyhook explanations. Second, altitude is not what distinguishes a revelatory event; it is distinguished by a pattern of significance, because only in the pattern can we hope to detect direction or a storyline. Dawkins is preoccupied with the height (the odds) of a few events (Fig. 14.1). Dennett, on the other hand, has essentially flattened out the landscape and asserts there are no fantastic coincidences and no events that defy what is imaginable (Fig. 14.2).[10] Dennett does allow internal cranes, such as the appearance of eukaryotic cells,

10. Although Dennett recognizes both figures as operative in natural selection, he seems to favor 14.2. This is evident in his debate with Gould about the importance of punctual equilibrium (*Darwin's Dangerous Idea,* 282–99).

sex, genes, memes, the advent of language, the human mind, and human culture. It is difficult to see a fundamental difference between a good trick and a skyhook, except that Dennett would say the former is from the ground up and the other is from the sky down; the former is imperceptibly gradual and the latter is dramatically visible. What if theologians accept Dennett's basic premise that there is one "fundamental process that developed the bacteria, the mammals, and *Homo sapiens*" (144)? I am not calling for a different mechanism, but for a different appraisal of a good trick along with the acknowledgement that Dennett's good trick is just as loaded and misleading as a skyhook—a good trick implies pure chance and a skyhook implies a heavy-handed creator. Both alternatives are false because they assume that chance and providence are mutually exclusive. In fact, no eye can see "behind the scenes" and identify one cause as divine and another as non-divine. Decisive events are what we must deal with.

My bare-bones discussion of such events names only a few that both stand out and have decisively changed the course of evolution or history (p. 231, n. 23). They are the pivotal nexus of whatever design the universe embodies. They are not to be dismissed so lightly as Dennett does when Johann Sebastian Bach is described as "an utterly idiosyncratic structure of cranes, made of cranes, made of cranes, made of cranes" (512). Dennett looks at the great edifice of evolution and sees one brick placed upon another brick but not the completed edifice.

Revelation includes an assumption that there is a divine intention that can be discerned—not by examining the cause of the event but by examining the pattern that consists of both the original "event(s)" and the ongoing incorporation of subsequent meanings arising from the original event(s). Not all events are revelatory, only those that disclose God's intention for the creation. Where do you look if that is your governing interest? The universe is rather large. Earth is much smaller but still rather diffuse. The conjunction of event and historical moment narrows the scope but the question remains: Which events? There are events that are paradigmatic and memorable but also monstrous and evil. The murder of Archbishop Oscar Romero reveals our dark side, but was it revelatory? To ask if there was a divine cause is to ask the wrong question; to ask if an event adds anything to the narrative truth of how we should live is the right question. The answer will come only from hindsight and then only as other events either make it worthy of remembering or disclose nothing of importance.

Jesus Christ as the Decisive Event

We speak of Jesus Christ as a decisive event because he meets the criteria of being both significant and revelatory. His life as a whole gathered up and projected into the future God's hopeful purpose for human life and the universe. His life galvanized the covenant made between God and Israel and the promise that Israel would be "a light unto the nations," if Israel would be faithful in its worship of God. His life was an event that projected a new future by extending the promise that Jesus Christ himself was "the light the darkness has not extinguished." This event deserves to be called revelatory because it disclosed a transcendent intention—an intention not of our own making and one climbing above its own historical moment. Leander Keck would have us remember that in some mysterious way, a transcendent event alerts us to the inadequacy of human language. We sense the ontological reality is inexhaustible. "The game is over," he writes, "if this Reality is not free to disclose itself through our language while at the same time breaking it sufficiently to inhibit our absolutizing it" (*The Church Confident*, 53). Its revelatory character is such that in spite of our limited thinking and speaking, the reality that grounds everything has a visible face.

Jesus Christ is a unique event insofar as his life, death, and resurrection were uniquely, but not exhaustively, transparent to the nature and purpose of the creator. The event of Jesus Christ was decisive because it was God's supreme act of self-expression—not the only event and not necessarily the last event, but the one event sufficient for human beings to be saved. In nonreligious language, the event of Jesus Christ was sufficient for human beings to have the opportunity to be lifted out of our self-destructive selfishness.

Christians have a definite belief concerning the *telos* of this unfolding. The essence of the plot of our story is found in the narrative of the ministry, deeds, and words of Jesus of Nazareth. A mystery that elicits, even overpowers, the religious self may and often does begin with a sense of wonder and awe. But this sense is not revelation except in the broadest sense, and does not become revelation "for me" until the individual lives a committed and faithful life patterned after a particular narrative truth. For the Jew and the Christian, that specific narrative truth is grounded in compassion and justice.

Once again, I emphasize the ontological and epistemological fit required to understand revelation. There is an objective and a subjective side to this revelation, as David Griffin argues, "a particular vision of reality which exists independently of our reception of it [its ontological significance] but is nevertheless made decisive for individual persons by our reception of it [its epistemological significance]."[11] On the ontological side, there are original events with original significance. The originality of an event has to do with its rootedness in a specific time and place.[12] Its meaning, however, resides in both its ontological givenness and in its epistemological appropriation. In the final analysis the ontology behind the name "God," determines what is an appropriate epistemology. There is something "beyond" and "beneath" the religious classic that makes it an event that transcends the past. Mark's

11. David Ray Griffin, *A Process Christology* (Lanham, Md.: University Press of America, 1990). I applaud Griffin's careful work to save revelation from the clutches of relativism. His book is a critical review of Paul Tillich, H. Richard Niebuhr, Rudolph Bultmann, and Friedrich Schleiermacher in terms of the crucial issue of the objective and subjective sides of revelation. On the ontological side, there are original events with original significance. This does not mean that they stand alone without epistemological appropriation or integration into a larger history or storyline.

12. For the importance of "the original" for revelation, see David H. Kelsey, *The Uses of Scripture in Recent Theology* (Philadelphia: Fortress, 1975).

gospel is a literary classic but its artistic style is not what makes it a *religious* classic. What makes its message unique is the dialectic it establishes between its author and ourselves, between the past and the present. Mark not only speaks of ultimate questions, he is grasped by them. In the Gospel of Mark, one finds few statements about God but a deep sense of being possessed by God in Jesus Christ. Those who read this gospel are connected to God because they are invited to accept their creatureliness over and against God's otherness. What makes Mark revelatory is the recreation of the event, Jesus Christ, in the reading (hearing) of the words that become transparent to the original subject behind them.

Conclusion

I hope we are finished with this need to identify revelation with cause and effect. No other motive has been more detrimental to theology than the attempt to make revelation more empirically acceptable to a scientific culture. The tangle gets worse when we try to identify God as the cause of some events but not of others. History is ripe with the diminishing returns of finding God at crucial unexplainable points. Newton represents a scientific tradition that required God as an ad hoc explanation (*deus ex machina*).[13] Others have tried to get around the inevitable difficulties of locating the divine in the distinction between primary and secondary causes. They have argued that God is the primary cause working through secondary causes.[14] Even among theologians there is widespread difficulty in keeping a balance between revelation as decisive and extraordinary without resorting to some form of supernaturalism or lapsing into mysticism. At one pole are theologies that are uneasy with supernaturalism but nevertheless speak of "a supernatural influence" and divinely revealed truth in verbal

13. See Richard S. Westfall, *Science and Religion in Seventeenth-Century England* (New Haven, Conn.: Yale University Press, 1958).

14. See Ian Barbour for a summary of all these approaches. I. Barbour, *Religion in an Age of Science* (San Francisco: Harper & Row, 1990), 117–18, 232–34, 244–50.

and propositional forms.[15] At the other pole are those theologies that so hide the divine action that it vanishes. Sallie McFague and Gordon Kaufman just about bury revelation in the sublime and the mysterious. God is the one who enlivens and empowers but does not order or direct. McFague writes, "Our focus is not on the purpose or direction of divine activity but on our dependence on God as the present and continuing creator" (*The Body of God,* 146). Kaufman will allow "genuinely theological patterns" and "cosmic trajectories" (proto-teleological directionality), but he is reluctant to identify specifically where God's purpose has been disclosed. Kaufman will speak of a particular divine direction only in the context of God's revelation in Jesus Christ, but it is not clear why this is the only "place" we can look to "deliberately and self-consciously" order our lives.[16] Consequently, for both theologians there is little or no reading of evolution or history as divine intention, only "the wonderful life" that has emerged from evolutionary history and with it our response of wonder, awe, and gratitude.

Contemporary theology is more sophisticated and subtle in how it speaks of revelation. It no longer is looking for a suitable crack to insert the divine. The shift is from top-down thinkers to bottom up. John Polkinghorne, a bottom-up thinker, speaks of "how God exercises providential interaction with creation" (*Belief in God in an Age of Science,* 62). Polkinghorne also writes, "I cannot give up the search for a causal joint" *(Belief in God,* 58). Perhaps that is the scientist speaking more than the theologian. There is common sense in what Austin Farrer wrote, "The grid of causal uniformity does not fit so tight upon natural processes to bar the influence of an overriding divine persuasion," and "in highly particular historical events, God from time to time shows his hand with a plainness the enlightened eye cannot mistake."[17]

15. I have in mind Carl F. H. Henry. See Gabriel Fackre's balanced assessment of Henry in his *The Doctrine of Revelation* (Grand Rapids, Mich.: Eerdmans, 1997), 154–75.

16. Kaufman, *In the Face of Mystery,* 389 and 486, n. 9, is as close as he comes to providing revelation with any ontological foundation.

17. The quote is from the excellent article by Brian Hebblethwaite, "Providence and

The enlightened mind will still need the heart of faith. It is here, where mind and heart join, that I will look to see the narrative truth about God.

This is the place to remember Wolfhart Pannenberg's double-sided thesis: The reality of God is co-given with our experience of the world, and God is accessible to theological reflection not directly but only indirectly. Therefore, revelation cannot be found in the direct cause of this event or that chain of events. Process theologians have rightly insisted that God is never the sole cause of any event, yet every event is dependent upon God for its existence. Because God is only indirectly co-given, and because no single thing or event is intelligible without its context of connecting events, the evidence for revelation is cumulative. Pannenberg tends to focus on the provisional and anticipatory nature of revelation, while I have emphasized the narrative character of the truth we seek about ourselves and the universe. "The reality of God," Pannenberg states, "is always present only in subjective anticipations of the totality of reality, in models of the totality of meaning presupposed in all particular experience."[18] It is possible for us to glimpse the meaning of our existence only if we are able to see patterns and directions as they unfold in human history and in the evolution of the universe. As Christians, we have a definite belief concerning the *telos* of this unfolding. The essence of the plot of our story is found in the narrative of the ministry, deeds, and words of Jesus of Nazareth.

It may very well be true that "God is the lure that arouses the cosmos toward adventure, constantly awakening it from the

---

Divine Action," *Religious Studies* 14 (June 1978): 223–36. The reference is to A. Farrer's book *Faith and Speculation*. See also Nancey Murphy, "Divine Action in the Natural Order," in *Chaos and Complexity: Scientific Perspectives on Divine Action*, ed. Robert J. Russell, N. Murphy, and Arthur Peacocke (Vatican City State and Berkeley: Vatican Observatory and Center for Theology and the Natural Sciences, 1995), 325–57.

18. Pannenberg, *Theology and the Philosophy of Science* (Philadelphia: Westminster, 1976), 310. I am uncertain what other "models" Pannenberg has in mind besides the life of Christ, because only in this event is the "totality of meaning presupposed in all particular experience" revealed. I wonder if Pannenberg means the totality of meaning or the essence (what really matters)?

inertia that would fix it into any given order,"[19] but no one is ever going to prove that by isolating a divine cause. The inquiry I would invite others to embark on is not one that reduces the natural world to one or several lucky, coincidental events. I would rather have the fuller story as it garners a richer account of the universe, our human nature, and who the creator might be. Even more intriguing is the web of events that incorporate chance and direction, contingency and design, the refractory nature of human actions and the inevitability of the divine plan (Paul Ricoeur, *Figuring the Sacred*, 182).

– SECTION 15 –
ISSUES OF CONTENTION AND PROMISE

Preliminary Observations

Are we on the verge of a " 'second phase' of the interaction between science and theology," as Arthur Peacocke suggests in his book cover review of *Rethinking Theology and Science* (1998)? The evidence in its favor is this: for the most part theology is recognized as a partner with important and novel cognitive claims. To this extent, we have moved beyond epistemic arrogance on the part of both participants with a simultaneous move toward mutual respect. An interdisciplinary space has been cleared that enables the recognition of shared resources.[20] Science and theology share some epistemic values, such as observation, argumentation, use of models or paradigms, and imagination. The outdated presumption that argumentation and imagination are particular to theology, while observation and verification are

19. John Haught, *The Cosmic Adventure*, 133

20. J. Wentzel van Huyssteen, *The Shaping of Rationality*. Van Huyssteen keeps pressing the argument that theology and science share "rich resources of rationality" and "epistemic values" (are these the same?); it is precisely at this point, where he generalizes, that he is vague. Included in the resources of rationality are: appropriate evidence, giving good reasons, critical standards, and performance rationality. These are not the same as epistemic values because when van Huyssteen speaks of these, he also mentions important differences. Compare pages 128 and 187 and my discussion in section 9.

solely the prerogative of science, is beginning to fade.[21] Closely related is the recognition that theology and science are beginning anew, epistemologically speaking, from a nonfoundational position. Characteristics of this shared postmodern condition include: the acknowledgement that experience and observation are "given" as already interpreted by language; epistemic values are shaped by the tradition and context in which they operate; criteria of acceptability are communal in nature; all knowledge is inherently contextual and historical; a plurality of voices and perspectives need to be heard and seen; and there is no God's-eye vantage point from which to see objectively and directly. One could also argue that the impossibility of a transcendent *methodology* means a transcendent *perspective* is only possible when traditions are challenged at their very core. In addition, scientists-theologians like Arthur Peacocke, Ian Barbour, and John Polkinghorne are sufficiently trained in both disciplines to give accurate accounts of science and technology. Lastly, the rise of two new disciplines, the history of science and the philosophy of science, have provided mediation techniques that greatly facilitate dialogue.

Nevertheless, it must be asked whether science and theology are mature enough for conversation. Maturity requires that both dialogue partners avoid stereotypes. Scientists violate this requirement more than theologians do. They see religious fanatics, fundamentalists, and creationists as the predominant voices. They often verge on simple derision. Prone to this charge are such well-known scientists as Richard Dawkins, Stephen Hawking, Francis Crick, and Carl Sagan. Hawking's famous quandary, "But if the universe is really self-contained, having no boundary or edge, it would have neither beginning nor end," naïvely concludes that

---

21. The era of positivism ended with the recognition that different contexts of meaning require different kinds of observation, evaluation, verification, and justification. Thus, while there are common epistemic values, the question remains whether there are significant differences arising from an empirical and a theological context of meaning.

there is no place for a creator (*A Brief History of Time,* 141). Whatever God did at the beginning is one part of the total narrative.[22] Scientists establish little more than a straw-man argument when they perpetuate God as a supernatural interventionist or a deistic God-of-the-gaps. There are theologians who look for God in the explanatory gaps, but many are more interested in the nature of the laws themselves as revealing God.[23] It is not a novel idea after all, that God works out divine purposes in and through the processes of the natural world.[24]

Maturity requires being less defensive and more open. Theologians have been on the defensive for so long, it is understandable that they resent any hint of science "lording it over them." Just as good science has little tolerance for theological answers to scientific questions, so good theology should have little tolerance for scientific answers to theological questions. But that is not the end of the discussion. Insofar as theology is doing realistic theology, it cannot afford to be indifferent to what science says about the nature of the world. Whenever science takes up the nature of being human, it cannot ignore a history of wisdom that is thousands of years old. Narrowly speaking, scientists can be good scientists without being theologically knowledgeable. This is less true, however, of the social sciences. One can be a theologian while being ignorant of science, but being so narrowly focused has its price. Theology is characteristically more pluralistic; to the extent that it bangs the drum of holism—and currently it is beating with both hands—it must accept the responsibility of listening to science, not as the voice of final authority but as one providing vital information. Ted Peters says simply and succinctly, "Sci-

---

22. See Kathryn Tanner, "Eschatology Without a Future?" in *The End of the World and the Ends of God: Science and Theology on Eschatology,* 222–37.

23. John C. Polkinghorne, "Creation and the Structure of the Physical World," *Theology Today* 44 (April 1987): 54. Polkinghorne writes: "God's activity in creation is not to be located with intervention in the world, either with or against the grain of physical world. Rather, it is to be found in those laws themselves, of which God is the guarantor."

24. For example, see Richard S. Westfall, *Science and Religion in Seventeenth-Century England.*

entific knowledge should inform and sharpen theological truth claims" (*Science and Theology: The New Consonance,* 1).

Another requirement for being conversational partners is that while participants may make specific claims for a particular methodology, this should not be interpreted as an attitude of superiority. Both science and theology are misrepresented when they are said to operate with a single, monolithic epistemology. My expectation is not for a common ontology or for a common epistemology, because that would be a mockery of some fundamental differences and would compromise the integrity of each. Both rivals need to honor the methodology of the other, each knowing that in some places there is consonance and in other respects they are seemingly irreconcilable.

Writing about interfaith dialogue, S. Mark Heim concludes that models that are inclusive or exclusive of truth claims are not very useful. He proposes instead the paradoxical rule that as one understands what is distinctive and unique about one's own tradition, one is then prepared to see what is crucial and unique in the witness of another religion. "To see nothing ultimately distinctive about one's own religious commitment in relation to others," he writes, "is as serious an impairment as to be ignorant of any religion but one's own."[25] Thus, theology and science should not be fearful or apprehensive about what is unique—or even exclusive—in their own claims. They should be eager to take hold of the differences to test them, compare them, and perhaps even have a change of heart and mind.

Two false choices emerge before us. One is to think of theology

---

25. S. Mark Heim, *Salvation—Truth and Difference in Religion* (Maryknoll, N.Y.: Orbis, 1997), 227. In his most recent book, *The Depth of the Riches: A Trinitarian Theology of Religious Ends* (Grand Rapids, Mich.: Eerdmans, 2001), Heim further emphasizes his conviction that by accepting different ends we allow for the mutual recognition of extensive and substantive truth in other traditions. In considering the interreligious dialogue, Harvey Cox urges participants to find vitality where their distinctiveness protrudes and rubs. His personal encounter with Asian theologies prompted Cox to ask of his own faith tradition: "Is there somewhere within Christianity itself, not just in its modern form, a defective gene that propels it into periodic outbursts of ruthless expansion, of crusades, of pogroms, and conquests, at the expense of other people's cultures?" See his *Many Mansions: A Christian Encounter with Other Faiths* (Boston: Beacon, 1988), 172.

and science as separate domains from which professionals communicate occasionally over philosophical bridges. The other is to mix and match scientific and theological insights and compose a symphony. These alternatives are usually seen for what they are—cheap and easy shortcuts that take us no place. It is possible that faith dispositions can be held in abeyance while scientists and theologians dialogue. The situation becomes more difficult when instead of dispositions we consider core beliefs. Often these core beliefs do not conflict because certain accommodations are made. In the issues below, the rub is felt when one's frame of reference (an epistemological-ontological fit) brings up stark philosophical and ethical choices. While I am sympathetic with the sentiment that neither science nor theology has the right to use a core belief as a trump card, I think Mark Heim's approach is better: core beliefs are neither tabled nor overruled; they are an essential element in what makes the conversation worthwhile.

In his introduction to *Science and Theology: The New Consonance,* Ted Peters proposes that God be treated as a hypothetical truth that is tested rather than assumed. He starts the conversation by asking if the God hypothesis "yields greater illumination—that is, greater explanatory adequacy—than working without this hypothesis" (1–2).[26] When God becomes a hypothesis, then God is not considered a core belief. In other words, for the sake of the conversation, can a core belief be treated as a hypothesis where "theological assertions about the absolute mystery we call 'God' become subject to critical evaluation in light of data gained from the natural sciences" (Peters, *The New Consonance,* 2)? If theologians, for the sake of the argument, are willing to treat God as a hypothesis, then scientists should be willing to treat naturalism or some other cherished core belief as hypothetical and ask if a theistic explanation yields greater illumination.

26. Similarly Walter Brueggemann, *Texts Under Negotiation,* 17; van Huyssteen, *Shaping of Rationality,* 258.

Dialogue requires the desire to learn from each other. In his *The Miracle of Dialogue*, Ruel L. Howel points out that communication becomes parasitical when the participants are not really committed to learn from the other, but instead simply select what confirms their own views.[27] Have scientists shown the necessary willingness to learn from theology? I ask this because it seems that the conversation thus far has been asymmetrical. Slowly emerging among theologians is the realization that science has been the dominant partner and that the conversation has been tilted toward making theological claims compatible with science. This is understandable as theology struggles to reclaim its own sources of authority. Repeatedly theologians have demonstrated that they know their science, many of them holding doctoral level degrees in scientific disciplines as well as in theology. The same cannot be said for scientists. Thus far, few scientists have shown much passion for the kind of conversation wherein theology is consulted for a fuller understanding of the character of the world we inhabit and the destiny of our planet. Historians and philosophers of science, while they have filled a critical breach, have not engaged theology in its classical form.[28] One must ask whether theology is to blame for this situation by withdrawing from the real world or whether the move toward realistic theologies has been too timid. Theologians must change the impression that they are "defenders of the faith" and not "seekers of truth." Scientists must start by being better acquainted with the writings of theologians, such as the ones quoted in this book, and by taking advanced degrees in theology as certain theologians have done in science. John Polkinghorne notes that there are scientists-theologians, such as Ian Barbour, Arthur Peacocke, Robert John Russell (not included in this survey), and himself, whose intellectual foundation was science; theologians who have paid particular attention to science, such as Wolfhart Pannenberg, Philip Hefner, Ted Peters, Nancey

---

27. Ruel L. Howe, *The Miracle of Dialogue* (New York: Seabury, 1965), 3 and 84.

28. Classical or original in the traditions of Saint Augustine, Thomas Aquinas, Martin Luther, John Calvin, and Karl Barth.

Murphy, and Thomas Torrance; and scientists who show an interest in exploring a theistic view of reality.[29] Michael Welker has observed that there can be a spirited exchange where scientists are eager "to engage some theological complexity" and theologians are willing "not to overrule scientific insights into reality by metaphysical presuppositions."[30]

Both theologians and scientists need to get their heads out of the intellectual ionosphere and wake up to the fact that decisions are being made every day that will shape our life on this planet for generations to come. Science is called upon to solve practical problems in a way that theology is not. Do I sense some hesitancy on the part of theologians to stick their noses in places where they are not exactly wanted? The same energy that has been devoted to cosmological issues should be spent on problems provoked by genetic engineering. Sallie McFague complains about energy misspent on the how and why questions and the lack of attention given to the more practical problem of "who God is, who we are, and what our responsibilities are as understood within this picture" (meaning the perennial questions, *The Body of God*, 77, 145). All too typical is the 1992 book, *Cosmos, Bios, Theos,* where sixty esteemed scientists responded to a few basic questions about origins, assuming that this theoretical issue is of great importance. My contention has not been that methodological questions are unimportant; I am concerned that they have been greatly overemphasized in the conversation. If the bending toward methodological questions was the consequence of theology's desire to be scientific or empirically successful, then the time has come for theologians to appreciate that they have a distinctive voice to offer.

When Michael Crichton introduced his bestseller *Jurassic Park* with these words, "The commercialization of molecular biology

29. See Polkinghorne, *Science and Theology* and *Faith, Science and Understanding* (New Haven, Conn.: Yale University Press, 2000), 155.

30. The occasion was an interdisciplinary group gathered at The Center of Theological Inquiry in Princeton in 1993. See Michael Welker, "God's Eternity, God's Temporality, and Trinitarian Theology," *Theology Today* 55 (October 1998): 318.

is the most stunning ethical event in the history of science, and it has happened with astonishing speed," he took notice of an entirely new phenomenon.[31] In every field of scientific endeavor, the almighty dollar looms large. If once sin was thought to be the curious mind unleashed,[32] it has become knowledge pursued with vested interests. Undoubtedly this has always been true to one degree or another. What has changed is the desire and potential to shape the future.[33] Biotechnology is only one way science will transform every aspect of human life: our entertainment, our food, our bodies, our health care, our environment, and our evolution. Crichton, with a note of alarm, comments that "it is done in secret, and in haste, and for profit," and "there are no detached observers, for everybody has a stake." We must pay attention to Jürgen Moltmann's observation that "if science and ethics are separated, ethics always appears too late on the scene" (*The Future of Creation,* 131). Let us remember that we have reached this point because forty odd years have been needed to remove barriers. Much effort has been expended to move both theology and science away from reductionism.[34]

---

31. Compare Bill McKibben, *The End of Nature* (London: Penguin, 1990). Here is a broader historical perspective where the contrast is made between a Darwinian age of survival and our era of "having our say" concerning who will survive.

32. See Blumenberg, *The Legitimacy of the Modern Age,* part 3 ("The Trial of Theoretical Curiosity").

33. Peter Berger was one of the first scholars to recognize the shift from a worldview where fate dominated to a worldview where choice rules. See his *The Heretical Imperative* (Garden City, N.Y.: Anchor/Doubleday, 1979). What has drastically changed over the intervening twenty years is that the potential has become possible.

34. Arthur R. Peacocke's *Creation and the World of Science* (1979) and *God and the New Biology* (1986), Ilya Prigogine and Isabelle Stengers, *Order Out of Chaos* (1984); or to overcome the dualism of natural versus supernatural: Teilhard de Chardin, *The Phenomenon of Man* (1955) and Alfred N. Whitehead, *Process and Reality* (1929); or to break the assumption that reason in science is opposed to faith in theology: Diogenes Allen, *Christian Belief in a Postmodern World* (1989), John Polkinghorne, *Reason and Reality* (1991); or the false choice between chance and providence: William Pollard, *Chance and Providence* (1981), John Polkinghorne, *Science and Providence: God's Interaction with the World* (1989); or the critique of foundationalism and objectivism: Michael Polanyi, *The Tacit Dimension* (1957) and Francis Schüssler Fiorenza, *Foundational Theology* (1984). As obstructions have come down, the constructive work of finding points of intersection increased. First came the use of models and paradigms (Ian Barbour, *Myths, Models, and Paradigms* [1974], Sallie McFague, *Metaphorical Theology* [1982]; Hans Küng and David Tracy, eds., *Paradigm Change in Theology* [1989]). Always in the background and then brought to the forefront were methodological parallels: Thomas F.

Issues of Contention and Promise

The conversation I urge begins with respect for the integrity of the differences in science and theology. In many respects, theology and science are two very different ways to frame meaning. The former is essentially historical, textual, experiential, and personal, while the latter is experimental, theoretical, observational, and impersonal. Both have developed different languages to bring to expression the ontology of a hidden creator or a verifiable universe. I have been working from the assumption that theology and science operate with different epistemological-ontological fits and that each has its own areas of specialized or preferred interests. I conclude this chapter with a brief look at those issues that not only enliven the scientific and theological inquiry but also remain obstacles in the way of a coherent narrative framework.

A Personal God

By tradition theologians defend belief in a personal God and in the kind of universe a triune God would create, while scientists by tradition find this belief to be inappropriate and not justified by the universe they know. This is such an obvious issue yet it is consistently sidestepped. I wonder if the confrontation is so long-standing that theologians and scientists assume there is no marketplace of exchange.

A core belief of the Judeo-Christian religion is a personal God who initiates, reveals, and enters into relationship. David Tracy, for example, writes in regard to revelation: "For the Jewish, Christian and Islamic traditions, this experience of the whole is an experience of a who; a loving and jealous, living, acting,

---

Torrance, *Theological Science* (1990), Nancey Murphy, *Theology in the Age of Scientific Reasoning* (1990). For a listing of centers devoted to research in the area of religion and science, see Ted Peters's introduction to Wolfhart Pannenberg, *Toward a Theology of Nature,* 5–6. The essays in *Science and Theology: The New Consonance,* also edited by Ted Peters, is representative of current trends and interests. For a good summary of where intersection between theology and science is being sought, see *Controlling Our Destinies: Historical, Philosophical, Ethical and Theological Perspectives on the Human Genome Project,* ed. Phillip R. Sloan (Notre Dame, Ind.: University of Notre Dame Press, 2000).

covenanting God, a God who discloses who God is, who we are, what history and nature, reality itself ultimately are" (*The Analogical Imagination,* 248). Before the discussion proceeds, a distinction must be made between one axis—let us call it the divine axis—and the cosmological axis. Now we can speak of a range from impersonal to personal along both axes, a gradient approach that removes the usual all-or-nothing choices.

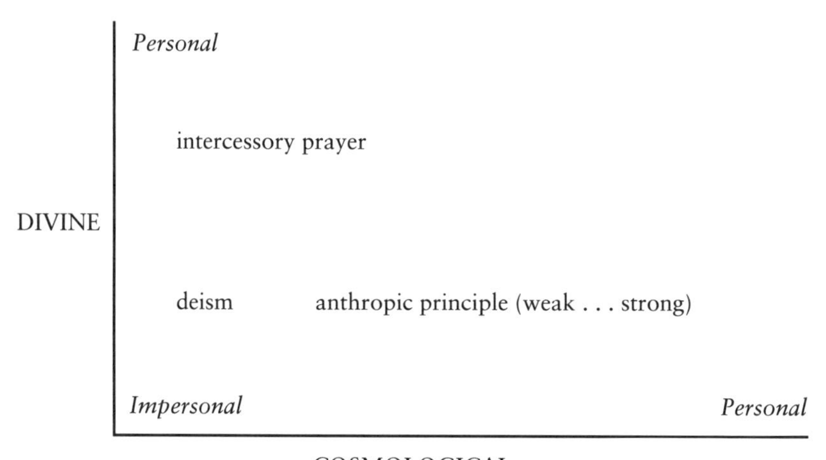

Figure 15.1

Does "personal" mean "a Being endowed with intelligence, will, affectivity, purposiveness, responsiveness" (Haught)? If we plot the attribute of purposefulness, the range would be from a universe that shows no indication of purpose to one where intentional activity abounds. On the divine axis the range might be from the creator who established universal parameters to a belief that God has a specific purpose for each human being. The trait of responsiveness has enormous breadth—from a deistic position to a belief in a God who takes a personal interest in my life and who will respond on my behalf. God's interceding can be hard (miraculous) or soft (a nudge). We can think of God along a spectrum from micromanager to macromanager. It is also true that the further we move along the personal axis, the further we move into belief dispositions that are not confirmable.

The highly respected Jewish theologian and philosopher Abraham J. Herschel wrote of a biblical ontology that understands God as one who has a passionate concern for the creation. This divine passion finds expression in God's personal and intimate relationship with the world and specifically in God's intention and will for human life. The central questions, then, become not when the universe came into existence but why it exists at all; not how it came into being but what is its fundamental nature.[35] Herschel did not have much success in moving the conversation away from its preoccupation with origins. Although Herschel encouraged an alternative ontology for God, moving away from an Aristotelian presumption where being predominates and God is the unmoved mover to an ontology of event and process where God is the most moved creator, scientists have been slow to appreciate the importance of the distinction. They instinctively treat belief in a personal God as taboo or as a philosophical sinkhole.

The key to the problem for both scientists and theologians is to reconcile a self-conscious and self-directing being with a universe that is at least ambiguous and "destructively ragged" (Polkinghorne). In his rejection of a personal God, theologian Gordon Kaufman draws an unwarranted conclusion usually associated with scientists. If one presumes a creator, it is difficult to avoid an understanding of God as one who is capable of self-conscious activity. It does not follow, however, that a self-conscious God must be pictured as a cosmic agent or as an intervening agent or as a superperson acting on behalf of personal requests.[36] We do not need to look for God as a distinct

---

35. See Abraham J. Herschel, *Between God and Man* (New York: Harper & Row, 1959) and *The Prophets* (New York: Harper & Row, 1962).

36. Gordon Kaufman, *In Face of Mystery,* 268–73. Likewise throughout Sallie McFague, *The Body of God.* Kaufman is undoubtedly correct in his assessment that a conception of God constructed on the model of a human purposive agent became associated with "a powerful teleological movement dominating the overall historical process" and that both have become "quite problematic in the twentieth century" (279). What does not follow is that why adopting a serendipitous reading of history, as Kaufman

causal factor in order to look for direction and divine intention. Whether we model God after ourselves is not a decisive argument either, since it is logical to find in ourselves qualities that reflect God's intention for creation. In the end it is the "how" of God's interaction with the world that has caused us to claim too much and know too little.

Newton, Darwin, and Einstein spent their careers trying to reconcile belief in a personal God with their own scientific descriptions of the universe. On the one hand, their quandary must be placed within the context of the hard ontology of God and the worldview they presumed. On the other hand, one asks how much of the quandary fades as the understanding of God and cosmos becomes more relational. Newton believed in a directing God though "He" was not very personal or biblical.[37] Darwin could not reconcile a personal, caring God with the capricious, "red tooth and claw" world of nature he encountered.[38] Einstein espoused a religion where God creates the universe in all its mystery and complexity but any moral implications are strictly derived from human beings left alone to their own devices and goodness.[39] What is seldom recognized is how a machine-like universe presented a different set of theological problems than did a capri-

---

does, should require jettisoning a personal understanding of God. I suspect a personal understanding of God is considered scientifically untenable and solely on that basis is unacceptable for anyone doing realistic theology.

37. For example, see Richard Westfall, *The Life of Isaac Newton* (Cambridge: Cambridge University Press, 1993), 119–29. This does not mean Newton was not a serious student of the Bible, including the patristic literature. Here he was as methodical and scientifically minded as he was in his pursuit of universal laws. If we make the distinction between information and formation, we would say Newton read the Bible as a sourcebook of information (perhaps mysterious information but nevertheless information to inform the mind rather than to form the heart). The contemporary shift is to believe that reading the Bible as information inherently distorts its message. In fairness to Newton, it must be said that he was an honest seeker of truth wherever it led him and in this instance he could not reconcile the orthodox doctrine of the Trinity with what he read in Scripture or in the universe.

38. See p. 237, n. 30. Desmond and Moore conclude: "Charles saw no salvation and suffered for it. Now he absolved himself on paper. As one with 'no assured and ever present belief in the existence of a personal God or of a future existence with retribution and reward,' he had not lived in fear of divine wrath. . . . 'I [Charles] believe that I have acted rightly in steadily following and devoting my life to science' " (*Darwin: The Life of a Tormented Evolutionist,* 636). In a real sense, science became Darwin's religion.

39. Max Jammer, *Einstein and Religion* (Princeton, N.J.: Princeton University Press,

cious, chance-driven model. In a machine-like universe, God is the grand designer and supreme mover. Here deism is the preferred scientific choice because God can be neatly removed from the picture. The biological paradigm presents a different set of questions because God cannot be so easily removed from the messy business of life. Darwin was troubled by the irreconcilable tension between the empirical fact of a capricious, purposeless world and God's omnipotence. One alternative that now presents itself is to understand the universe in a more relational or feminine manner. Although science is moving in this direction, it remains firmly entrenched as a masculine operation. Because as a discipline theology is discovering its feminine side, does theology then become a needed counterbalance to a masculine science?[40]

Intercessory prayer becomes a test case, par excellence. Chet Raymo, professor of physics and astronomy at Stonehill College in Massachusetts, speaks for many naturalists when he describes prayer as "a meditation on the world, informed by knowledge, open to mystery." In his book of "natural prayers," such prayers "are answered not with miracles, tagged with our names or those of loved ones, but with beauty and terror." All of his life, Raymo confesses, is a "letting go of incantational magic, petition, and the vain repetition 'Me, Lord, me' in order to be more attentive to the light that burns at the center of every star, every cell, every living creature, every human heart" (*Natural Prayers,* Introduction). John Polkinghorne, on the other hand, is not willing to relinquish a place for intercessory prayer. Prayer is the human response whereby we come to know God's intention. Prayer, then, is the aligning of two wills.[41] Here is a serious difference. Polkinghorne is able to envision God as caring about each individual (Matt.

1999) and Thomas Torrance, "Einstein and God," in *Reflections,* a publication of the Center of Theological Inquiry (spring 1998): 2–25.

40. What is decidedly different for theology is its legacy of women theologians, such as Julian of Norwich, the impact contemporary women theologians such as Rosemary Radford Ruether are having, the increasing number of women entering the ministry, and, of course, the mother Church.

41. Lecture given by John Polkinghorne at Harvard University Chapel in 1992 and *Belief in God in an Age of Science,* chap 3.

10:29–31) without chaining it to the capricious intervention of God (Why me and not you? Compare Deut. 11:13–15). Raymo is mistaken if he thinks the Bible is a story where "we are swept along on the grand wings of an abiding plan and presence" (*Natural Prayers,* xiv), for the salvation story it tells is not about "all things bright and beautiful" but about the despised and the rejected who are loved by their creator. The pertinent theological question is whether a scientific description of the universe is inherently at odds with the Judeo-Christian story of a redeeming God who enters into all of life as it has evolved.

Providence is the other critical test case. This book is an example of how to reinterpret providence without doing violence to any fundamental scientific principles. Stated more positively, I have argued that providence is essentially the storylike character of life on this planet as it is moved forward by decisive revelatory events. My purpose is not to argue for a natural theology in the strong sense of presenting evidence from the natural order to prove the existence of a personal God. Rather, I wish to take seriously a scientific view of the world, knowing that it is limited and embedded in its own interests, and ask how it coheres with a theological understanding, likewise circumscribed and driven by its own governing interests.

In the final analysis, we must decide if we and the universe were made for love. While each moment may be a struggle for survival, this fact can coincide very well with the other fact that the struggle has resulted in creatures with the capacity for knowledge of the transcendent. William Blake held this tension together in his immortal "fearful symmetry" of the hand that made both the lamb and the tiger.

Methodological Questions

One sign of maturity on the part of both science and theology would be the recognition that on a significant level they operate with distinctive methodologies. This does not deny other methodological contacts, such as intelligibility, aptness, coherence, warranted assertibility, falsification, verification, consensus building, probable reasoning, and critical realism. The following arguments have followed us throughout the course of this book. In summary form they are:

Religious Truth	***Modern or Scientific Truth***
Transcendent subject	A reality to be manipulated
Ontological priority	Procedural priority
Passive reception	Active manipulation
Disclosed	Constructed
Meditation	Discovery
Hearing/obeying	Theorizing/experimenting
Inspiration	Thought experiments

The methodological standoff is mitigated by the fact that theology is a rational construction of the beliefs of the Christian community. That is, theology is a second-order description of a first-order, highly personal experience of the other. The moment the experience of the divine is brought to expression, it enters the public sphere of rational-critical discourse. The crux of the difficulty is how theological assertions submit to the criteria of falsification. From a scientific viewpoint, theology is hobbled by a methodology that is circular, even though such a methodology is not necessarily pernicious or irrational. The circularity is the result of having no starting point that is independent of the frame of reference that is being tested.[42] Concerning the crucial perennial question about the nature of being human, for instance, the Gospels would say that we have no idea what the human condition is before we meet Jesus. Old Testament writers would claim that we have no idea what the universe is until we know its creator and so forth. If the starting point is faith (faith seeking understanding)[43] and faith is a gift from God, there is no place to begin that is hypothetical. Here is a fit between ontology and epistemology

42. In his classic book *The Structure of Science* (1961), Ernest Nagel defines a scientific law as one where the physically real "can be identified in ways other than, and independent of, the procedures used to define those things" (144).

43. See Karl Schmitz-Moormann, *Theology of Creation in an Evolutionary World* (Cleveland: Pilgrim, 1997), chap. 1.

where the methodology cannot be defined until the ontological reference is known. I have not forgotten that I have argued that there is a certain circularity or reciprocity in doing science, because the procedure shapes what is to be known and something other than the knowing self shapes the methodology. However, the similarity does not go beyond saying that theology and science are playing different games in the same stadium.

A range of methodological issues could be considered here, but I am particularly interested in just one that is fundamental. The standard scientific methodological presumption is naturalism—only natural cause and effect relationships are considered and allowed. But why should theistic assumptions be excluded a priori? One could assume a world of intentional causes as well as one where there are no divine causes. Thus, one could ask if it is necessarily better science to exclude theistic possibilities or whether it is necessarily better theology to include them. Interestingly enough, in *The Blind Watchmaker* Richard Dawkins readily admits the biologist's problem is how to explain complexity and design, which "is highly unlikely to have been acquired by random chance alone" (9). His solution to the problem of the origin of life is to evoke natural selection as the blind watchmaker. But why is this necessarily the best inference or the hypothesis with the greatest explanatory power? Both naturalism and theism are philosophical assumptions about the way the world is. Why is one legitimate and the other not? Why isn't Dawkins "called to the bar" for allowing his presumption of methodological naturalism to cloud his conclusions, which surely would be the case if he began with theistic assumptions?[44]

Perhaps the critical test is not simply explanatory power or degeneration. The scientist would have us consider whether one methodology is more resistant to revision and includes beliefs that

---

44. For a conservative approach that asks these very questions, see Howard J. Van Till, *Portraits of Creation: Biblical and Scientific Perspectives on the World's Formation* (Grand Rapids, Mich.: Eerdmans, 1990) and Paul de Vries, "Naturalism in the Natural Sciences: A Christian Perspective," *Christian Scholar's Review* 15 (1986): 388–96.

are not open to falsification. If the answer is affirmative, then it is bad science; one could argue that it is bad theology as well.[45] On this level, there is no decisive argument. Both naturalism and theism harbor methodological beliefs that are not only assumed true but also highly unfalsifiable. When Michael Ruse asks, "Is evolution a social construction?" he gives both an affirmative and a negative answer because the universe is filled with phenomena (mysteries) that have such excesses of meaning that beliefs and "meta-values" invariably enter into the explanation.[46]

The problem with doing science with theistic assumptions has been the inclination, if not the intention, to discover gaps, perplexities, sudden emergences, or irreducible complexities that are best explained by evoking intentional causes or a designer. Let us remember that in the eighteenth and nineteenth centuries scientists waged a battle to get theology out of science for precisely the reason that by evoking a creator you have not explained anything scientifically.[47] Most scientists do not want to fight that battle again. When explanations are intermingled with belief dispositions, those belief dispositions can operate in two ways: as governing interest and/or as controlling principles.[48] Being aware of those dispositions should be the most important methodological concern because, as Wolterstorff reminds us, everyone has a religion in the sense of something that provides a unifying coherence to his or her endeavors. It is for this reason that most scientists and theologians would prefer to see as clean a separation as possible between beliefs and empirical hypotheses. Wearing more than one hat at the same time is asking for method-

---

45. It could be so argued but not necessarily. We have returned to the issue if, in the conversation with scientists, theologians must or should suspend their core beliefs. See how this is developed in Mark Heim's book (see n. 25).

46. Michael Ruse, *Mystery of Mysteries* (Cambridge, Mass.: Harvard University Press, 1999), especially chap. 12.

47. Stephen Gould would have us ask: If it took centuries for science to rid itself of unwanted and unwarranted theological questions, why would science invite them back in now? The proper response would be: very cautiously.

48. Nicholas Wolterstorff, *Until Justice and Peace Embrace*, 166–72.

ological confusion. For the most part, it is the individual and not the discipline that attempts larger and more comprehensive accounts. I am thinking specifically of physicists like Niels Bohr and Werner Heisenberg; philosophers like Alfred North Whitehead and Teilhard de Chardin; poets and humanists like Loren Eiseley and Jacob Bronowski; biologists such as Erwin Schrödinger, Theodosius Dobzhansky, and Ludwig von Bertalanffy; historians such as Herbert Butterfield and Stanley Jaki; philosophers of science such as Michael Polanyi and Stephen Toulmin. What distinguishes these minds was not that they turned to philosophy or theology as an escape from their own disciplines, but that they did so in order to assess the meaning of what they had found. Mutual respect and mutual challenge happens at the level of belief dispositions, which are in various degrees implicit in doing science and theology. It is theology's place to ask science to account for why the universe seems to be filled with purposeful activity. This kind of question stretches science and exposes its own methodological limitations without declaring that methodology wrong. Science, on the other hand, confronts theology with its implicit anthropomorphic, one-planet deity.

What Is and Who God Is

Let us assume that the way the universe *is* informs us about the way its creator *is,* and the way the creator *is* informs us how and why the universe *is.* If, speaking hypothetically, the ontology of God can reveal the true nature of the universe and the ontology of the universe can disclose the nature of God,[49] then there is something of substance to discuss. Let us also assume that theology and science come to know reality from different methodologies

49. The language I use here is important for I did not say the ontology of the universe, as it is known by science, can disclose the *true* nature of God. When "true" takes on the meaning of "full" or "complete," then Christian theology, with one voice, affirms that revelation is needed to truly know the nature of God. Christian theology would also insist the true nature of the universe cannot be known apart from revelation, but science can disclose true traits of the universe that are not accessible through a strictly theological methodology.

and that their epistemologies have been shaped by many factors, not the least being the realities they encounter and their governing interests. That science does not necessarily discover God, or that theology does find "signals of transcendence" (Peter Berger), is just as symptomatic of methodological preference than of reality itself.

The evidence of the last several centuries shows that theology is likely to adopt the ontology science posits as foundational for an understanding of the universe. This constitutes a reversal from the many centuries when theology was queen and the nature of the universe was read through doctrines concerning the nature of God. This is not the place for a long historical digression on this trajectory, but notice what has happened as science matured into its modern form. Because Kepler assumed a coincidence between word (theological) truth and empirical truth, he looked for theological significance in the truths to be discovered about the cosmos. The true nature of both God and the cosmos coincided and could not do otherwise since there is only one creator.[50] Darwin, however, pursed his science independent of theological propositions, and did so in such a rigorous and consistent manner that he set a new standard. Darwin effectively shut the book of nature and thereby killed the belief in natural theology so prevalent in preceding centuries. Theologians, most emphatically neo-orthodox theologians, also rejected natural theology on methodological grounds. The consequence has been a theology of nature, where who God is and what the universe is are set in dynamic tension.[51]

---

50. Kepler did not exactly use Scripture as a guidebook to how the universe "works." This bar against moving easily from Scripture to universe was even more pronounced with Galileo, because he was forced to think through the methodological implications of empiricism vis à vis Scripture.

51. See Eugene Thomas Long, ed., *Prospects for Natural Theology* (Washington, D.C.: Catholic University of America Press, 1992). Pannenberg, for instance, will argue that "God is a factor in defining what nature is, and to ignore this fact leaves us with something less than fully adequate explanation of things," but does not want to merge or diminish the critical edge inherent in two separate disciplines. See his *Toward a Theology of Nature*.

What Is and What Should Be

There is a pervasive and long-standing attitude that it is highly fallacious to link "what is" and "what should be."[52] Albert Einstein held such a view in his classic definition of the different roles science and theology should have: the former to explain what is and the latter to interpret what should be done.[53] This same judgment is echoed clearly by Stephen Gould's "denouncement of the misguided search for intrinsic meaning within nature" (*Rock of Ages,* 178). The reluctance to read values off anyone's ontology is justified to the extent that ontologies have consistently been manipulated to become a means of subjugation. We should know by now how easily the "ought" of one person or one group is given ontological justification in order to shield it from suspicion or criticism.[54]

In *The Cosmic Adventure*, John F. Haught finds it more satisfying to defend a cosmic purpose that embodies beauty rather than one that points to an accession of self-consciousness (á la Teilhard de Chardin). Haught's argument that beauty and adventure are values that can be read off nature is an important one, but it too is open to criticism. Stephen Gould calls this the fallacy of "all things bright and beautiful" or you can't get there (God) from here (nature is all warm and fuzzy and orderly).[55] When Carol Newsom asks if the natural world is a locus of intrinsic value,

---

52. Alasdair MacIntyre's *After Virtue* was instrumental in demonstrating that many schools of philosophy since Hume and Kant declared that no "ought" can be derived from an "is." See Charles Taylor, "Justice after Virtue," in *After MacIntyre*, for an illuminating discussion of MacIntyre and why that demonstration is fallacious (16–23).

53. Jammer, *Einstein and Religion,* 135.

54. Gould should be credited with exposing the many instances where science has been misused to justify unwarranted ethical conclusions. For example, see Gould, *Bully for Brontosaurus* (New York: W. W. Norton, 1991), 316–18; *The Panda's Thumb* (New York: W. W. Norton, 1980), chap. 4. Jeremy Rifkin gives this issue considerable attention in his *Algeny* (New York: Viking, 1983). For example, he writes, "Every civilization justifies its behavior by claiming to have natural order on its side" (35).

55. Gould speaks of the fallacy of "all things bright and beautiful" as a stellar example of a type of theological thinking that must be confronted if "advocates wish to argue that the moral meaning of life lies exposed in nature's factuality." Gould, though, has only one line of reasoning and so he cannot get to God via "all things bright and beautiful." Almost without exception, evolutionary biologists know nothing of a suffering God or a kenotic theology (see below).

she reminds us that Christians see the cosmos as full of purposive activity, affirm the intrinsic goodness of world, and encounter a certain order and rightness inherent in creation.[56] In her study of *Job*, she finds theological significance in a world that is populated and is itself wild, untamable, even chaotic. "What kind of ethic begins in contemplation of such as this?" she asks (23). It is an ethic that begins with the world as genuinely other: a cosmos that Job did not create, cannot dominate, control, or tame for his own use. By not making light of Job's sense that he has been treated unjustly, the story does not press the moral sense of the universe as if it is an intrinsically good universe justifying an intrinsically good God. Newsom writes, "God populates Job's world with beings [Job 38–41] that he has never had to take into account as genuine others. Job's previous organizing metaphors were those of mastery and dependence drawn from the social world of village patriarch. Now he is presented with creatures whom he must understand in relations of nondependence and nondomination" (25). The ethic one can build upon is this: Job is not God but a creature; nevertheless, a creature capable of understanding better than the rest of the creatures a world that is more than monotony, triviality, chaos, and senseless suffering.[57]

Holmes Rolston echoes this sentiment and carries it a step further by stating that nature is cruciform. Like Polkinghorne, Rolston understands contingency (chance, chaos, and natural selection) as necessary by-products of a universe that exercises a freedom characteristic of how God creates.[58] But contingency is

---

56. Carol Newsom, "The Moral Sense of Nature: Ethics in the Light of God's Speech to Job," *Princeton Seminary Bulletin* 15 (New Series 1994): 9–27. See also the significant arguments made by Daniel Migliore for a created universe reflecting its creator. Migliore, *Faith Seeking Understanding* (Grand Rapids, Mich.: Eerdmans, 1991), chap. 5 ("The Good Creation").

57. Newsom makes an observation I have missed. After learning how fragile life can be, Job participates in the goodness of creation by giving birth to ten more children and "finally and most intriguingly, he gave his daughters an inheritance along with his sons, a gesture that would lessen their dependence in a patriarchal world" (Job 42:15; Newsom, 26). It would be unfair to say that Newsom is only concerned about a moral ethic. She reads off nature an ontology that speaks to the very ontological nature of God; namely, a Yahweh who cannot be domesticated and used.

58. John Polkinghorne, *Belief in God*, where he writes: "They [the nature of dense snow

more than a necessary feature of life that allows for the emergence of something higher. "The pathetic element in nature," Rolston writes, "is seen in faith to be at the deepest logical level the pathos in God. . . . The secret of life is seen now to lie not so much in the heredity molecules, not so much in natural selection and the survival of the fittest, not so much in life's information, cybernetic learning. The secret of life is that it is a passion play."[59]

This "is" of nature is more than the beginning of an ethic because it reveals the very nature of God as one who is the ultimate suffering Being who empties "her" very self into creation. Rolston's cruciform ontology does nothing to exclude God as the benevolent architect but decidedly includes God as the suffering redeemer. No one should overlook the historical shift that has taken place. As Isaac Newton "read" the universe, God became the great designer. The capricious ways of nature were simply not acknowledged. Darwin not only acknowledged contingency, he made it the cornerstone of his worldview. But what about the capricious nature of human beings? Postmodern Christian and Jewish theologians feel that in order to do theology with any degree of integrity they are required to face the horrors of World War II and the Holocaust and to do so without letting go of belief in a personal, caring God.[60] Can the scientific community, likewise, say its integrity is somehow dependent upon facing up to the unprecedented death and destruction of the modern, empirical era?

Erazim Kohák has no hesitancy about recovering the moral sense of humanity by recovering the moral sense of nature. His *The Embers and the Stars* is a powerful statement that we have

---

fields to sometimes slip with destructive force, etc.] are the necessary cost of a creation given by its Creator the freedom to be itself. Not all that happens is in accordance with God's will because God has stood back, making metaphysical room for creaturely action" (13).

59. Holmes Rolston III, "Does Nature Need to Be Redeemed?" *Zygon* 29 (June 1994): 205–29.

60. See Richard Rubenstein, *After Auschwitz* (Indianapolis: Bobbs-Merrill, 1966); Elie Wiesel, *A Jew Today* (New York: Random House, 1978); Darrell J. Fasching, *Narrative Theology after Auschwitz* (Minneapolis: Fortress, 1992); Emil Fackenheim, *To Mend the World: Foundations of Post-Holocaust Thought* (New York: Schocken Books, 1989).

sought the sense of life where it cannot be found: in products of our technology and in empirical arguments: "Thus my primary tool has been the metaphor, not the argument, and the product of my labors is not a doctrine but an invitation to look and to see" (xiii). The moral sense he would have us rediscover is to relearn the rhythm of day and night, the beauty of solitude, the experience of nature as a meaningful whole. The latter is found when we experience the community of life and joy, work and rest, pain and healing that binds all of us together on this planet.

When we seek to learn our role in the universe by looking to the universe, we are in the proper place to address the third perennial question. We could ask it as Jeremy Rifkin raises it: "But how would we know what the cosmos expected of us?" (*Algeny*, 248). Rifkin answers his own question by concluding that we must cooperate with the community of life rather than exploiting it for our own security and survival. He assumes that we owe something to the cosmos in return for everything we have taken from it (a sense of indebtedness). When he writes in his last chapter that our species should represent the interests of the cosmos, I become queasy with the move to define our role vis à vis the "good steward."[61] To think of the earth as "home" and "community" is the better way to define our place in the universe and our role as creatures who bear the burden and blessing of transcendence. Jürgen Moltmann describes home as a place where there is a network of relaxed social relations. I am at home where people know me,

61. In forging a new ecological ethic, Mark I. Wallace urges theologians to "abandon the repugnant rhetoric of protection and stewardship and substitute in its place a new language of humility and caution." Our dubious role as "benign exercisers of dominion" should give way to a role "as dangerous travelers on a fragile earth—an earth that has had enough of our enlightened oversights as it cries out for us to leave it alone before it is too late." Wallace, "The Wild Bird Who Heals: Recovering the Spirit in Nature," *Theology Today* 50 (April 1993): 25–26. Stewardship in the best sense observes the limits inherent in the order of nature, but it does not sufficiently guard against our abiding inclination to overreach those limits. The good steward does not necessarily look to nature as a teacher. We are beings who are so very young—that is reason enough to let this vast universe be our teacher.

where I am acknowledged without fighting for recognition.[62] Our attitude toward nature in this regard has been ambivalent. On the one hand, there is the Western attitude that not until nature has been shaped into an environment does it becomes a home for us. The early American settlers felt obliged to tame the West and clear the forest before they called it home. The native population, however, had learned how to walk lightly and work with nature. Calvin Luther Martin makes a more radical indictment. Martin believes that much harm has been done with the "invention" of history where a chosen people, priests and kings, laid hands on the cycle of time and straightened it into a vector in order to hurl their mandate, "their trajectory, in truth their history, against the agents of evil, error, and folly" (*Spirit of the Earth,* chap. 3). We are not that far from Rifkin's own indictment that we feel obliged, at least in the West, to define our place by laying hands on nature, remaking it, and then sanctifying our statue with a worldview. What frustrates community is not simply death, decay, and selfishness. We are heirs of an age of urbanization and industrialization that has deeply altered and narrowed our sense of community.

In the biblical vision of Shalom, this sense of frustration is overcome when nature is able to enjoy nature ("the wolf shall dwell with the lamb"), when human beings enjoy nature ("they shall plant vineyards and eat their fruit"), and when nations will benefit each other ("and they shall beat their swords into plowshares" (Isa. 11:6, 65:21; Mic. 4:3). Nicholas Wolterstorff places the emphasis on enjoyment: "Shalom at its highest is enjoyment in one's relationships. . . . To dwell in Shalom is to enjoy living before God, to enjoy living in one's physical surroundings, to enjoy living with one's fellows, to enjoy life with oneself" (*Until Justice and Peace Embrace,* 70–71).

One of the several common threads binding together the Hebrew and Christian Scriptures is the sense of belonging to a single

62. Jürgen Moltmann, "The Alienation and Liberation of Nature," in *On Nature,* ed. Leroy Rounder (Notre Dame, Ind.: University of Notre Dame Press, 1984), 138–39.

community with many voices and levels of self-consciousness capable of praising our creator.[63]

> Praise Yahweh from the heavens!
> praise him in the heights.
> praise him, all his angels,
> praise him, all his hosts! . . .
>
> Praise Yahweh from the earth,
> sea monsters and all the depths,
> fire and hail, snow and mist,
> storm-winds that obey his word, . . .
>
> kings of the earth and all nations,
> princes and all judges on earth,
> young men and girls,
> old people and children together.
> (Psalm 148, New Jerusalem Bible)

Humans: Nothing Special, Just One among Many

Theology and science have such different responses to the third perennial question concerning our place in the universe because they do not see eye to eye regarding the question regarding our human nature. It is axiomatic among scientists that empiricism does not allow the human species any stature beyond that of animal.[64] Stephen Gould is typical: "*Homo sapiens* may be the brainiest species of all, but we represent only a tiny twig, grown but yesterday on a single branch of the richly arborescent bush of life. This bush features no preferred direction of

63. Nature in fact can show human beings the way, since "this song of praise was sung before the appearance of human beings, is sung outside the sphere of human beings, and will be sung even after human beings have—perhaps—disappeared from this planet." See Jürgen Moltmann, *God in Creation* (San Francisco: Harper & Row, 1985), 197.

64. It is difficult to know whether this judgment is made on purely methodological grounds or if ideological reasons have imposed themselves. Michael Ruse points out that scientists, evolutionary biologists in particular, inevitably import meta-values into their conclusions. These are values that shape and justify the context of the science rather than actually entering into the science as content itself (*Mysteries*, 143). However, if the former happens, can science be methodologically pure regarding its content?

growth, while our own relatively small limb of vertebrates ranks only as one among many, not even as *primus inter pares*" (*Rock of Ages*, 180).

Theology's starting point is the tradition of the *imago Dei* ("made in the image of God") and the human soul.[65] Scientists typically dislike this hypothesis because it implies that humans are unique since they are endowed with certain qualities (meta-values) "from above." This is not necessarily so, however, because a nondualistic picture of being human is exactly what is argued in *Whatever Happened to the Soul?* If there is one value and one characteristic that has traditionally distinguished the human person, it has been the soul. In this volume (*Whatever Happened to the Soul?*) and from various disciplines, including biology, psychology, genetics, neuroscience, philosophy, and theology, the soul is not an entity separable from the body. Warren S. Brown summarizes: "Thus 'soul' (or at times 'soulful' or 'soulish') is used herein to designate not an essence apart from the physical self, but the net sum of those encounters in which embodied humans relate to and commune with God (who is spirit) or with one another in a manner that reaches deeply into the essence of our creaturely, historical, and communal selves."[66] What we find is an effort to redirect the concept of soul as a distinct "thing" and to relocate it in the capacity for a particular realm of experience. There are several ways to name that realm of experience—the experience of the holy, personal relatedness, the capacity to enter into a covenantal relationship with God—but all of them declare that we are more than our genes. Thus we find that some theologians will concede that nearly all, perhaps even all, the human abilities traditionally

65. I am not equating soul with *imago Dei*, because the latter refers to our *role* in the created order (covenantal partner with God), while the soul is the appropriate way to speak of the *capacities* we have in order to fulfill that role (Gen. 1:26–27). As creatures made in the image of God we are called to fulfill the role of caretakers ("dominion over") because we have the capacities for transcendence (to pray and to know we are creatures of the creator) as well as language, faith, hope, and love.

66. Warren S. Brown, "Cognitive Contribution to Soul," in *Whatever Happened to the Soul?* ed. Warren S. Brown, Nancey Murphy, and H. Newton Malony (Minneapolis: Fortress, 1998), 101.

attributed to the soul (and heart) can be understood as functions of the brain (language, emotion, behavior, and decision making). What remains uncompromised is the emergent (sum total) capacity to seek and know the intention of the creator. For me, that one distinguishing characteristic of human beings is that we alone pray.[67]

Are humans capable of transcending their environment, their genes? If we have evolved to serve the interests of our genes, then all religious efforts to motivate, inspire, and sustain individuals in faith communities are deceptive and doomed to failure. There is more than just a little reductionism in the thesis that the individual is only capable of acting in one way: to protect and reproduce genetic material. Richard Lewontin, the noted evolutionary biologist, urges a more holistic understanding of development. Lewontin emphasizes that he is not offering anything that biologists do not already know. What he finds missing is an explanatory narrative that should incorporate the organism and the environment as both cause and effect in a co-evolutionary process (*The Triple Helix,* 126). In living organisms the boundaries between internal and external are soft, allowing for mutual influence.

To respond to the third perennial question, science finds itself in an awkward place. Theology may also experience a different kind of awkwardness because it can be only so reductive until the vertical dimension of the divine is so hidden and emasculated that it is indistinguishable from neuroscience or psychology. Some theologians feel that line is being crossed in *Whatever Happened to the Soul?* Because human beings are so like the other creatures, science, on the other hand, cannot assign them any exceptional responsibilities—yet that is exactly what scientists are apt to do. Stephen Gould, like many biologists, urges upon our species the responsibilities befitting a superior kind. He is typi-

---

67. In his systematic theology Robert Jenson understands humans to be distinguished from the other creatures because we alone pray. See his *Systematic Theology,* vol. 2 (Oxford: Oxford University Press, 1999), 58–59.

cal in his comment, "We are therefore left with no alternative. We must undertake the hardest of all journeys by ourselves: the search for meaning in a place both maximally impenetrable and closest to home—within our own frail beings" (*Rock of Ages,* 177).[68] Charles Taylor is certainly correct in identifying the shift that has taken place beginning with the Enlightenment. As the ontology of the cosmic order was displaced by the epistemology of right procedures, our lives could no longer be ordered by a vision of order located outside ourselves (transcendent). Instead, "the source of obligation is... rather my own status as a sovereign reasoning being, which demands that I can achieve rational control. I owe this, as it were, not to the order of things, but to my own dignity."[69] Is it not ironic that we should look within ourselves when we are not even *primus inter pares* (first among equals)? In a neutral world observed by a disengaged, free, rational self, where is the "good" to be found? There is nothing left but the disengaged, free, rational self and Descartes's "I am" is beset by the onslaught of doubt that constitutes the postmodern condition. And if there is no scientific or philosophical ground on which to stand to know what is good or how to respond to the third perennial question about our role in the universe, then could it be that theology offers an entirely different way to proceed?

---

68. Steven Weinberg's nihilism is well known ("The more the universe seems comprehensible, the more it also seem pointless"). There is some point, however, to the search itself. "But if there is no solace in the fruits of our research, there is at least some consolation in the research itself. Men and women are not content to comfort themselves with tales of god and giants, or to confine their thoughts to the daily affairs of life; they also build telescopes and satellites and accelerators.... The effort to understand the universe is one of the few things that lifts human life a little above the level of farce, and gives it some of the grace of tragedy." Weinberg, *The First Three Minutes* (New York: Basic Books, 1977), 154–55.

69. Charles Taylor, in *After MacIntyre,* 19. Early on, E. F. Schumacher had a clear grasp of our epistemological-ontological crisis. Schumacher saw the crisis as the result of a turn to an all-knowing and self-reliant self. The ensuing crisis in epistemology was the embracing of a progress that simultaneously excluded wisdom from epistemology. This, Schumacher writes, "was something we could perhaps get away with for a little while, as long as we were relatively unsuccessful; but now that we have become very successful, the problem of spiritual and moral truth moves into the central position" (33). Seeing things as they really are is not just a matter of getting the equations right, because a person driven by greed or power loses the ability of "seeing things in their roundness and wholeness" (31). See E. F. Schumacher, *Small Is Beautiful* (New York: Harper & Row, 1973).

Authority

Theology and science have different sources of authority that permeate both their epistemologies and their ontologies. Over the centuries, these sources and uses of authority have been in no small measure the underlying crux of their rivalry.

I have already referred to Jeffrey Stout and Alasdair MacIntyre's highlighting the dilemma inherent in theology. With theology's domain greatly eroded, there was an initial retreat to a reasonable religion and then to religious experience and more recently, as I have argued, to right methodology. Stout writes, for instance, "Only by restoring to Christian theism its cognitive dimension and by taking seriously precisely those paradoxical doctrines that make it seem distinctive to secular thought could it retain a role of any centrality in modern culture" (*Flight from Authority,* 139). If theology must make the same kind of cognitive claims as science in order to be taken seriously, then it must downplay the paradoxical character of its statements and abandon reference to revelation as the basis of its authority to speak truth. If theology follows this course, it sacrifices what makes its voice distinctive and likewise authoritative. Theology has been confronted with the accusation that its standard bearers—such as sacred writings, history, and tradition—can no longer be taken for granted as self-authenticating. In contrast, science rose in stature by the success of a methodology that was self-consciously not self-authenticating.

Does this mean that in order to be taken seriously theology will have to demonstrate sources of authority that are empirically falsifiable? Nancey Murphy made a valiant effort to this effect, but for reasons already discussed, it was like forcing a square peg into a round hole. In order to translate the historical and dialectical character of theological truths into empirical-like statements, the very heart and soul of Christian truth is forfeited. In spite of the laudable and valid efforts to find common methodological meeting places, science and theology are methodologically divergent. The difference diminishes as both engage in the work of organizing comprehensive frameworks. At the same time, however, the potential for rivalry increases, because at the level of meta-values opposing and irreconcilable presumptions surface. There are beliefs and values, as this section illustrates, where the task of integrating narrative truths around perennial questions becomes difficult.

Nancey Murphy is aware of the reductionism that is necessary when working at the level of testable hypothesis. The unexamined question is whether this is the level that is appropriate for theology. It would seem that it is not, since Murphy herself has joined the chorus of voices urging a more holist epistemology and, more specifically, a holistic justification.[70] Unfortunately, the nature of this holistic justification is strictly intratextual, that is, theology is understood as primarily descriptive of Christian belief and practice with the criteria of acceptability residing within the Christian community and the biblical text. As viewed from the outside by scientists, this holism, where "belief and meaning are inextricably related" (Murphy) and words merely reference other words, sounds like more of the same holy speculation that is not grounded in anything real or testable.

I believe there is a general agreement that theology must continue to regard Scripture as its primary source of authority, even if a variety of ways exist to understand and use that authority.[71] Unfortunately, within the ranks of theologians there is no more fiercely debated issue than this one. It continues to divide Christians along a liberal-conservative demarcation. Two kinds of solutions have been proposed. Liberal theologians like Paul Tillich and David Tracy appeal to a critical correlation between Christian texts and common human experience (*Blessed Rage for Order,* chap. 3). Thus, they keep one historical standard but introduce the universal role of human experience with the expectation

---

70. Murphy, *Theology in the Age of Scientific Reasoning,* 201–3.

71. See David Kelsey, *Uses of Scripture in Recent Theology* (Philadelphia: Fortress, 1975), especially chaps. 4 and 8.

that this constitutes a methodology of scientific verification. The conservative position as defended by Carl F. H. Henry and the Southern Baptist Convention understands Scripture as an objective norm providing propositional truths.[72] This puts Christian theology at a serious disadvantage, because its most important authoritative source is plagued with controversy and weakened by disagreements.

What happens when theology recommits itself to word-truth embedded within narrative truth? Truth as word both speaks to and arises out of human experience and history. But the conversation with science must go beyond the word-truth of experience, history, and Scripture and speak to the ontology of the physically real of the universe. Theology and science will need to be more attentive to the distinctive claims, fits, and types of verification that accompany truths that are explanatory, descriptive, cognitive, and performative. The theological fit has always been characterized by a tight weave of the cognitive and performative aspects of truths—performative in the sense that religious truths are pressed into practical use, telling us how to live life. Science, to the contrary, works diligently to treat explanation and verification as separate procedures in order to maximize objectivity. The word-truths of theology are ultimately driven by the ontology of a hidden and self-disclosing Subject.[73] The empirical truths of science depend upon an ontology of "objects" that can be quantifiable, measurable, and explainable.

What is arguable is whether belief and meaning are interrelated in the same way for theology and science. There is some veracity behind the notion that empirical truths derive their meaning from

72. See George Hunsinger's helpful comparison of Carl Henry and Hans Frei. Hunsinger, *Disruptive Grace: Studies in the Theology of Karl Barth* (Grand Rapids, Mich.: Eerdmans, 2000), 338–60. Hunsinger correctly points out that Carl Henry is not the best resource for developing a conservative position that has the potential of bridging the breach with liberals. There are other conservative theologians, such as Abraham Kuyper, Herman Bavinck, and more specifically, G. C. Berkouwer in his *Holy Scripture* (Eerdmans, 1975) who understand the difficulties of using the Bible as an objective norm.

73. See William Stacy Johnson, *The Mystery of God: Karl Barth and the Postmodern Condition* (Louisville: Westminster John Knox, 1997).

the things in the world they explain, while theological truths receive their meaning from questions that go beyond the real. Words are authoritative for scientists when they form testable hypotheses that reference universal activities. The authority of words for theologians is dependent upon their faithfulness to the human experience and the tradition of words that are considered revealed. If some new ontological reality were discovered, words would be used empirically to explain the phenomenon as like or different from other phenomena, while words would be used theologically to extrapolate its meaning as illustrative of a larger theological understanding. The fuller significance of the new reality would result from two different sources of authority rather than from their forced conjunction.

The issue can be clarified further. Revelation and inspiration point to the phenomenon that human words do on occasion have a quality of transcendence. Theology cannot claim that God speaks in divinely inspired sentences and propositions, but it can claim to have such words that have the potential to mediate truth about the nature and will of God. Or, to express it differently: Scripture is inspired because the Spirit enabled members of a community in a particular time and place to articulate what there was about a particular configuration of events that made it uniquely significant for the salvation of the world.[74] The writer or speaker is "overshadowed" by the Spirit (Luke 1:53). As I have argued throughout this book, theological truths do not depend upon the inerrancy of the words themselves but on the narratives they form. The inspiration is indicated in that they reveal something about human nature that no human wants to hear or can even hear without faith and obedience. If inspiration means anything, it means that ordinary language is transposed in such a way to disclose a new dimension of reality. Science is primarily a reaching out to calculate, grasp, manipulate, and understand reality. Theology, on the other hand, must listen first for another

74. Colin Gunton, "All Scripture Is Inspired...?" *Princeton Seminary Bulletin* 14 (1993): 249.

word, not its own, and then speak as one who has been grasped, used, and reborn.[75]

The Transcendent Hermeneutic versus the Reductionism Principle

We begin by remembering that science drove out the principle of transcendence from its domain because it undermined the empirical process. When theology brings transcendence into the discussion, it is tantamount to talking about God, and God is not a verifiable datum. Remember Daniel Dennett's contrast between "skyhooks" and "cranes"? It is a further refinement of the standard scientific exclusion of anything that looks or smells like a miracle. The movement of evolution by cranes is allowable because it refers to natural selection and incremental changes. Skyhook, on the other hand, is shorthand for miracles or the dramatic, sudden emergence of some perfection, such as the eye.

Transcendence is an explanatory principle to which theologians are entitled but not in the usual sense. The usual sense is to factor in a divine action. What theologians should reject is the use of transcendence to explain what is for the moment unexplainable and the use of transcendence as short hand for "God did it." What does this eliminate? It eliminates any and all searches for divine causes as if they can be isolated and identified. What does this leave? It leaves space for an explanation that is holistic in character. Transcendence is the thesis that in order to understand the complexity of life and the universe, higher-level explanations are necessary and are different from the summary of the parts. Transcendence, then, is the counter-principle to reductionism. Transcendence is not to be evoked under the pretense of being an empirical-like explanation. Nor is it a fill-in-the-gap explanation. While it borders on faith and is associated with cer-

75. John Dominic Crossan captures this insight that there are questions we cannot ask ourselves when he clarifies why Jesus' use of parables is different: "It is clear that parable is really a *story event* and not just a story. One can tell oneself stories but not parables." See *The Dark Interval* (Allen, Tex.: Argus Communications, 1975), 87.

tain faith commitments, it is not to be judged as an irrational or a subjective response. A mature science should be able to move past its condescending attitude that transcendent explanations will inevitably plunge us into an abyss from which there is no escape.[76]

The real rub with science is the theological argument that the supervenience of higher-level explanations is the only adequate explanation that allows for free will, conscious intention, personal responsibility, and ethical decisions. The same issue is fiercely debated concerning emergence. Arthur Peacocke defines emergence as "the entirely neutral name for that general feature of natural processes wherein complex structures, especially in living organisms, develop distinctively new capabilities and functions at levels of greater complexity."[77] This "new, non-reducible reality" is at the same time both entirely "natural" and ontologically distinct. Thus Peacocke, as well as the contributors to *Whatever Happened to the Soul?* argue a "new, non-reducible reality" gradually made its appearance with the advent of the human species. This new level of life transcends all that went before but is not discontinuous with what preceded it and prepared the way for it.[78] This is said in contrast to Pope John Paul II's assertion that the soul must have been "immediately created by God."[79] Ernan McMullin, the much-respected philosopher and historian of science, observes, "[W]hen so much weight is placed on non-reducibility as the criterion of this 'new reality,' it would clearly be

---

76. As a hermeneutical principle, transcendence looks and sounds very much like the boundary questions David Tracy described under the larger category of "limit experience." See his *Blessed Rage for Order* (Minneapolis: Winston/Seabury, 1975). "I hope to show how, at the limit of both the scientific and the moral enterprises, there inevitably emerge questions to which a response properly described as religious is appropriate" (92).

77. Arthur Peacocke, "Relating Genetics to Theology on the Map of Scientific Knowledge," in *Controlling Our Destinies,* 347.

78. For another discussion of emergence, see *Whatever Happened to the Soul?,* 101–3. What is not asked: How can a new level of life emerge (transcend) that constitutes a new ontological reality when it is "not discontinuous with what preceded it?" This is the question scientists ask and where theologians usually resort to skyhook language.

79. See Pope John Paul II, "Message to the Pontifical Academy of Sciences on Evolution," *Origins* 26 (November 1996): 350–52.

important to establish that the distinctively human capacities are, in fact, non-reducible."[80] Here is where transcendence and emergence reach stalemate between theologians and scientists. The latter has an aversion to ontological leaps and new levels of being, so liberal theologians have smoothed out the discontinuities and done away with heavenly cranes. For most scientists there is no unambiguous ontological distinction between the human and the nonhuman. Theologians maintain there are distinctively human features that are irreducible. We have, thank goodness, moved away from trying to identify some genetic mutation or some point in the hominid line when God infused the first soul(s). But the issues of transcendence, emergence, and irreducibility will not go away—and they should not.

Theologians have their own quandary to ponder. While embracing the physicalism of the universe and offering theological explanations that do not contradict scientific statements, the theological principle of transcendence has itself gone through a reductive process. If transcendence does not imply a divine agency that exists outside of time and space, it has lost one of its essential meanings.[81]

Evil and Sin

Why don't scientists speak about evil and sin more often? Is it because these terms are vacuous or without substance? Is it because they are theological in nature? Are they so tainted and burdened with tradition and misuse that they are irredeemably damaged? Those who choose to view the universe from a nontheistic point of view will have to explain history—what we have

80. Ernan McMullin, "Biology and the Theology of the Human," in *Controlling Our Destinies*, 370–71.

81. Compare Huston Smith's classic theological definition of transcendence: "By transcendence I mean something superior to us by every measure of value we know and something that eludes us." Smith then comments: "To expect a transcendental object to appear on a viewing screen wired by an epistemology that is set for control [empiricism] would be tantamount to expecting color to appear on a television screen that was built for black and white." See his *Beyond the Post-Modern Mind* (New York: Crossroad, 1982), 114.

done to our own species, to every other species, and to the health of our planet. If this is not sin, then what is it and why has it not dissipated as our knowledge and self-consciousness increase? Without some ontological grounding, I am afraid sin will forever be nothing more than what human beings do, as mistakes that can be corrected.

In contrast to an understanding that depends upon psychological, sociological, or historical notions, theologians define sin theocentrically. From a theocentric perspective, sin is against God; it is what humans willfully do against the will of God. This makes little sense unless the universe is viewed theocentrically, as created by a self-conscious, intentional God. Although creation is not the only conceptual scheme that can frame sin, it remains the best model in order to draw a sharp line between our finitude and God's infinitude.[82] The God-creature conceptual map is essential to how we understand ourselves and our situation. Gordon Kaufman writes, "To posit God, to believe in God, to commit ourselves to God, is to acknowledge that we are not absolute, not autonomous: we are finite, limited, subject to error, failure, self-delusion, corruption; and therefore, our willing, our agency, our defining of what we should be and do, our ideas of right and wrong, must always be regarded as questionable" (*In Face of Mystery,* 368). Without this theocentric perspective, it surely needs to be asked: What is to limit this human creature that desires to know without limits?

An ontological understanding of sin thoroughly rejects the presumption that sin is a moral deficiency that can be corrected. Edward O. Wilson sees nothing naïve in his proposal "to invent moral reason of a new and more powerful kind," since he holds firmly to the Enlightenment belief in the advancement of moral

---

82. For other conceptual homes for sin see David H. Kelsey, "Whatever Happened to the Doctrine of Sin?" *Theology Today* 50 (July 1993): 169–78. Kelsey argues that sin has not so much disappeared as migrated to other trajectories, such as anthropology. Andrew Delbanco offers a very different account of how the sense of evil has been lost by Americans. See his *The Death of Satan* (New York: Farrar, Straus & Giroux, 1995).

reasoning.[83] If sin is rooted in our capacity for transcendence (to reach beyond ourselves), as I have argued, then it will forever be our shadow self in whatever we do, even the moral good we intend.

Theology will challenge science to acknowledge sin and evil as realities that are as real as any particle or universal principle. Science, on the other hand, will call into question those doctrines of evil that make nature the culprit. In this conversation we must distinguish between sin and evil. Sin refers to our willful acts while evil is the ocean of selfishness and violence from which sin drinks. Walter Wink gives evil added depth when he speaks of it in terms of "some irreducible power which cannot finally be humanized, cured, or integrated, but only held at bay."[84] In the historical experience of being human, every individual makes the evil that is already present a new reality. Original sin may have something to do with a genetic disposition to overreach, but it has more to do with the evil that already predates everyone's birth. There is a tendency to equate the anteriority of evil with the cosmic structures of the universe or an inherited human flaw. David Ray Griffin is justified in his denouncement of "natural evil," as if nature erupts or loses control of itself willfully (*God, Power, and Evil,* 275–310). Why are natural catastrophes evil when they are part and parcel of nature's autonomy and sovereignty? The structures of the universe do not know sin, nor is evil an appropriate way to understand the nature of nature.

As we write about the narrative truth of the universe, it will be more than interesting to see how sin is treated and incorporated. We enter an era when our desire to know all things grows exponentially and, like the Christian doctrine of sin, becomes ever more critical. Sin was very much at the heart of the biblical story of sin and redemption. Even though the physical picture of our

---

83. See Edward O. Wilson, *Biophilia* (Cambridge, Mass.: Harvard University Press, 1984), 119, 139, and passim.

84. Walter Wink, *Unmasking the Powers* (Philadelphia: Fortress, 1986), 28. For a more subtle and powerful treatment of evil in a science fiction story, see Robert Reed, *Marrow* (New York: Tom Doherty Associates, 1997).

story has changed dramatically, the ontological predicament of being human has not.

Finished versus Unfinished/Long versus Short Story

One of the hidden edges of the rivalry between science and theology is the tension between plurality and the certainty as they arise from the perspective each uses to tell the narrative story of the universe.

Religion has many stories to tell and many interpreters of these stories, all having a sense of ultimacy. This sense of ultimacy and certainty comes from claims of revelation and inspiration. For Christians, it also comes from a sense that the essential story has been told. Unlike the scientist who has a vast universe to search, the theologian knows where to look. This is the *proleptic* character of revelation, which discloses the "end" of history. The end, so to speak, has already been anticipated in the Buddha or in Jesus Christ or in the history of God's covenant with Israel. The opportune moment (*kairos*) is so intense that Jesus Christ is "the condensation of History into a single dot of ultimate significance, the condensation of time into reality" (Ellul, *The Humiliation of the Word*, 251). Because theologians operate from the assumption that the meaning of human existence has already been decisively revealed, their perspective is one of certainty and completion.

If the cosmos is like an immense jigsaw puzzle, scientists work to fit together the seemingly infinite number of pieces with no end in sight, since the edges continue to expand and change.[85] They work to complete the picture where there are no edge pieces. Yet scientists have a firmer grasp of the "short future." Their narrative style may be episodic but it too has a ring of certainty since there are laws and principles that are universally applicable. However, the certainty is of a different quality: the difference

---

85. Whether the universe is changing substantially (ontologically) is one of the great unsettled questions in science. As a layperson, it seems to me that many unsettled questions of considerable consequence remain, which leads me to understand science as beginning and not nearing the end of its work.

between optimism and hope (see below). If the theological perspective is panoramic, the scientific "take" is a series of clearly focused snapshots. This means we have two storylines that are not easily integrated.

The irony of the situation is that science is on surer ground epistemologically speaking because the scientific community is able to build consensus based on methodological consistency. Science appears to have a methodologically firm foundation when its ontological subject, the universe, is equivocal beyond words. On the other hand, Christian theology does not possess the same methodological certainty and yet claims to know the essential character of the human story. Clearly the religions of the world cannot speak from consensus because methodological diversity and disunity exist, but that has not stopped them from speaking with unequivocal authority about their ontological subject.

In *Science as Salvation* Mary Midgley writes, "Science is important for exactly the same reason that the study of history or of language is important—because we are beings that need in general to *understand* the world in which we live. . . . All human beings need some kind of mental map to show them the structure of the world" (33). For the most part theology has chosen to narrate our story about our place in the universe utilizing word-truth. This means it exercises the freedom of a poet's eye for metaphor and a prophet's cry for justice.[86] Scientists easily become preoccupied with the smallest details over the longest expanse of time; consequently, they tell a fragmented and disjointed story. Often science's narrative explanation is grayish and anemic. Edward Wilson's *Consilience: The Unity of Knowledge* is a good example because religion is shorn of any explanatory power and natural science is afforded all rights to intellectual coherence and moral insight. This splitting of the intellectual scene has made mapmaking much harder, as many have observed.[87] The pur-

---

86. I am thinking in particular of Walter Brueggemann, *Finally Comes the Poet: Daring Speech for Proclamation* (Minneapolis: Fortress Press, 1989).

87. Midgley, *Science and Salvation* (New York: Routledge, 1992), 37. See also Bryan

pose of this book is to find common ground in narrative truth, where the tension between an empirical and a theological perspective is sustained by the integrity of each. The differences in their points of view provide the promise of a denser and more colorful story.

Optimism versus Hope

Science is inherently optimistic while Christian faith is fundamentally hopeful. We can press this comparison still further by asking: What is the ultimate hope of each? Idealistically, scientists hope for and work toward world peace, the end of hunger and disease, a more complete understanding of the universe. These are worthy endeavors shared by almost everyone. What brings these hopes into the domain of science is the optimism that by human endeavor they can be realized. They become "microhopes" we can predict because they are based upon a confidence that with the proper application of scientific expertise the world can be made new. It would be unfair to say that science does not inspire hope. It is, however, a hope on a short tether. Science inspires optimism because it has no doubt that it will take us to the next frontier (the very warning of Crichton's *Jurassic Park*). And so we "advance" by grasping one rope after another, but each rope is suspended in midair.

Janet Soskice speaks to the core characteristic of Christian hope when she writes that hope, like faith and love, *abides* and endures "even in the midst of profound evil, and without ignoring that profound evil" (*The End of the World,* 78). Likewise Fraser Watts comments, "Indeed, hope characteristically occurs in situations of darkness or uncertainty in which optimism would be impossible or out of place" (*The End of the World,* 57). Hope abides because it is not tethered to a visible or predictable future. Its teleological end is not the rapture or the transformation of the earth. Jesus'

Appleyard, *Understanding the Present: Modern Science and the Soul of Man* (New York: Doubleday, 1992) and Heinz R. Pagels, *The Dreams of Reason* (New York: Bantam, 1988).

resurrection bears cosmic significance because it protests the law that entropy is the last word. The incessant decay of matter is not the "sting" that throws us into the darkness of cosmic futility. God's ultimate purpose is not even dependent upon humankind or planet Earth. That would render God rather small and human beings rather gigantic. Christian hope abides because there is the particular faith that nothing can separate us from the love of God, including things present and things to come, life and death, powers and principalities, nor anything else in all creation (Rom. 8:38–39).[88]

Why is it that the same science that expounds a cosmic futility seems to be immune from cosmic despair? Richard Dawkins begins his book *Unweaving the Rainbow* with this quote from his colleague Peter Atkins:

> We are the children of chaos, and the deep structure of change is decay. At root, there is only corruption, and the unstemmable tide of chaos. Gone is purpose; all that is left is direction. This is the bleakness we have to accept as we peer deeply and dispassionately into the heart of the universe.

Dawkins lauds this kind of tough-mindedness and proper purging of saccharine false purpose, but then reflects: "Presumably there is indeed no purpose in the ultimate fate of the cosmos, but do any of us really tie our life's hopes to the ultimate fate of the cosmos anyway? Of course we don't; not if we are sane. Our lives are ruled by all sorts of closer, warmer, human ambitions and perception."[89] Science can afford to rebuff beginnings and endings as inconsequential because it is teleologically bankrupt, which means that it can only enjoin local or short narratives.[90]

---

88. Douglas F. Ottati argues for a "hopeful realism" that is opposed to both "easy optimism and cynical pessimism." His hopeful realism and his theological construction resonate well with the realism I am proposing. See Ottati, *Hopeful Realism: Reclaiming the Poetry of Theology* (Cleveland: Pilgrim, 1999).

89. Dawkins, *Unweaving the Rainbow: Science, Delusion and the Appetite for Wonder* (Boston: Houghton Mifflin, 1998), ix–x.

90. Jean-François Lyotard described the postmodern condition as being able to have

Once you eliminate teleological purpose, not much of any story can be told. The narrative that science does manage to tell is one of evolution blindly groping, punctuated by emergent probabilities. Beginnings and endings become important only when there are connections to be made. Science provides important raw material for the story about the universe and our place in it, but science's particular epistemological-ontological fit is habitually prone to exclude many of the perennial questions necessary for a full understanding of the universe and our place in it.

What constitutes hope is ambiguous. Many hope for material gain. Others hope to accrue technological knowledge. Everyone hopes for a better tomorrow. What will we become as human beings if our hope is wrapped in an optimism that denies sin and discounts evil? It matters greatly how our story begins and how we think it will end; we dare not believe that science and theology tell the same story. Christopher Lasch, that astute political observer and historian, notes that believers in hope think of themselves as the party of hope when actually they have little need of hope. "Hope does not demand a belief in progress," he writes. "It demands a belief in justice."[91]

Conclusion

The conversation between science and theology is about making a meaningful narrative of both the human and the cosmic story by paying attention to all three perennial questions. This is undoubtedly a difficult undertaking, but the urge to do it is as ancient as human reflection itself. Theology and science do not lock horns or complement each other often because they are each engaged in their own particular interests, but when they do, the stakes are high. Because meaning is not an explicit property of reality, the

---

only little and local narratives locked in combat. See his *The Postmodern Condition: A Report on Knowledge* (Minneapolis: University of Minnesota Press, 1984).

91. Lash, *The True and Only Heaven: Progress and Its Critics* (New York: W. W. Norton, 1991), 80–81.

human mind must discover, construct, imagine, remember, theorize, philosophize, and theologize. Science has demonstrated the value of an empirical way to know the universe. Theology, on the other hand, draws upon a reservoir of wisdom and oral reflection. Over the centuries, theology has demonstrated the value of words to guide, motivate, compose, create, inquire, and transcend. Because reality is both discovered and encountered, it imposes certain restraints both on how we come to know it and on how we can represent it by using words or mathematical symbols. As human beings we experience reality to be intelligible, interesting, and beautiful not simply because we are intelligent, interesting, and beautiful ourselves, but because the universe is all of these things in and of itself. As our understanding of the universe increases, we come to better know ourselves, our place in the universe, and the possibility of the creator.

Conversation between these two siblings in the twenty-first century will need to keep in mind theology's temptation to indulge in ungrounded speculation and pretentious claims and science's predisposition toward reductionism and myopia. Behind theology's love of word-truth lies the beguiling attraction to construct clear and certain doctrines from thought alone. The Enlightenment was the era when we believed we could "reason through" to God's purpose. Inspiration became confused with intuition and natural law was a facile reading of the eternal laws of God's mind. In this conversation, science needs to be reminded of its prideful optimism. In a *Time* magazine essay entitled "In Search of the Silver Bullet," Charles Krauthammer writes:

> There is something touching and terribly human about our belief in the powers of science. Its roots, of course, are deeper and older than science. The longing to penetrate the impenetrable is ancient. Satisfying that longing was once the mission assigned to magic, then religion. As these have declined, the task has evolved upon science. Science has much to offer the world in machinery and ideas. But it is too much

to ask of it (and too little to ask of ourselves) to discover the key to murder and genius and to distinguish truth from lies. (Nov. 4, 1985)

Science has done more than compel theology to ask new methodological questions. It has created a new picture of the world that radically alters how we ask the perennial questions. The great Judeo-Christian themes of creation, salvation, and providence cannot be told as if the narrative contexts have not been drastically altered. As science expands the edges of the universe and "humbles our place in it," how do we understand God's acts of redemption on our planet? The centrality of Christ in God's plan of redemption is not only challenged by interreligious dialogue but by a dramatically revised perspective. Sallie McFague speaks of the "scandal of uniqueness" as it is absolutized by Christian claims that God is embodied in one place and only one place. McFague argues that such claims seem skewed, to say the least, when set against the fifteen-billion-year history of the universe with its hundred billion galaxies (*The Body of God*, 159).

As we debate the justice of artificial insemination or abortion or cloning, we find no information about genetics in the Bible. Its view of the world is circumscribed and limited. Its cosmology is wrong. The secularization of public discourse is a fact and theology severely limits its role in shaping the future if it does not heed the world as it is described by science. It is nearly unthinkable, John Polkinghorne complains, that Jürgen Moltmann's theology of creation does not consider seriously the role of chance as treated by quantum physics, general relativity, thermodynamics, and evolution.[92] To the degree the criticism is valid, it cuts both ways, of course. Not every theological treatment of creation must engage science or vice versa; neither should one discipline be ignorant of the other. Once the treatment of creation or the

92. See John Polkinghorne, *Science and Creation*, 2, referring to Moltmann's *God in Creation* (1985).

universe slides into a metaphysical worldview, then engagement is desirable and necessary. In his provocative article, "Does Nature Need to Be Redeemed?" Holmes Rolston reminds us that the language of Bible and biology are not entirely unrelated, "for they do each offer a concept of nature, a worldview." He continues, "Both biology and Bible do seek to characterize nature and, even if one does so scientifically and one poetically, the two descriptions need to be congenial."[93] Why do they need to be congenial? We might hope there would be places for crosstrafficking, but we should equally expect them to be rather discordant as is the case with John Haught's theological treatment of evolution.[94]

Surely one task of theology is to focus on how life is to be lived and how reality is to be construed "in the light of God's character as an agent as this is depicted in the stories of Israel and the life of Jesus" (Lindbeck, *The Nature of Doctrine,* 121). But how to construe reality is also a primary concern of science. Christian theology invites the accusation of self-description and even self-deception when its criteria for truth are solely intersubjective and intertextual. This might be the kind of criteria appropriate for a tradition of word-truth—except this word-truth includes ontological truth claims concerning the nature of reality. With so much emphasis on rescuing theology from her epistemological isolation, we have overlooked the long-standing suspicion that it is soft on reality. Can theology slough off the substantial undermining of Marx, Feuerbach, Freud, Kierkegaard, Foucault, Dawkins, and Hawking? What reality are we looking at? As Edward Farley reminds us, "in the Christian master narrative, faith takes place in situations of human struggle, suffering, evil, and sin. It is a certain way of living toward God and toward the world while existing in the midst of such things" (*Deep Symbols,* 70). Faith then, as

---

93. Rolston, *Zygon* 29 (June 1994): 207–8.

94. Haught's *God after Darwin: A Theology of Evolution* (Boulder, Colo.: Westview, 2000) is based upon the premise that the model for conversation is neither opposition nor separation but engagement. This allows for some sharp edges of disagreement to remain exposed. Haught does what Moltmann does not do (n. 90), namely, *engage* science from a theological perspective.

Farley asserts, turns human beings toward and not away from the world and it is a world of crosses and heartbreak and not all things bright and beautiful. In our anxiousness to find points of contact in the mystery and wonder of life and the universe, there is the eerie silence that differentiates theology and science when the real is the tragic, inexplicable, and ugly.[95]

Ultimately, the question of meaning is a human question more than it is a scientific or a theological one. We are possessors of, and possessed by, a fundamental human spirit that reaches out to penetrate the essential nature of our universe. It may be that the disjunction between the human self and the external world is the insidious beginning of metaphysical speculation. Nevertheless, it is a beginning that cannot be avoided. The most salient feature of the human species is its capacity to transcend the immediate and to aspire to infinity. Since the beginning, human beings have felt a pressing need to narrate truths of capacious and significant proportion. What those truths are and how they define our place in the universe, theology and science, joined with every other human endeavor, will strive so to discover and tell.

---

95. The use of the word "ugly" is meant to be provocative. It is not only theologians who only see what is good in what is beautiful. Scientists have their own proclivity toward what is symmetrical and perfect and thereby avert their eyes from what is malevolent, fractured, and downright repulsive. On the scientific side, see Timothy Ferris, *Coming of Age in the Milky Way,* chap. 16 ("Rumors of Perfection"). Science, however, is learning the importance of the asymmetrical, but this is about as close as it comes to approaching the ugly. See Frank Close, *Lucifer's Legacy: The Meaning of Asymmetry* (Oxford: Oxford University Press, 2000).

Bibliography

Addinall, Peter. *Philosophy and Biblical Interpretation.* Cambridge: Cambridge University Press, 1991.

Appleyard, Bryan. *Understanding the Present: Science and the Soul of Modern Man.* New York: Doubleday, 1992.

Bailey, George. *Galileo's Children: Science, Sakharow, and the Power of the State.* New York: Arcade, 1990.

Barbour, Ian G. *Issues in Science and Religion.* New York: Harper Torchbooks, 1996.

———. *Religion in an Age of Science.* San Francisco: Harper & Row, 1990.

Barr, James. *The Concept of Biblical Theology: An Old Testament Perspective.* Minneapolis: Fortress, 1999.

Barrett, William. *The Illusion of Technique.* Garden City, N.Y.: Anchor, 1979.

Barth, Karl. *Church Dogmatics.* Translated by G. W. Bromiley and T. F. Torrance. Vol. III/2, *The Doctrine of Creation.* Edinburgh: T. & T. Clark, 1960.

Baumer, Franklin L. *Modern European Thought: Continuity and Changes in Ideas, 1600–1950.* New York: Macmillan, 1977.

Becker, Carl L. *The Heavenly City of the Eighteenth-Century Philosophers.* New Haven, Conn.: Yale University Press, 1963.

Becker, Ernest. *The Denial of Death.* New York: Free Press/Macmillan, 1973.

Becker, Jasper. *Hungry Ghosts: Mao's Secret Famine.* New York: Free Press, 1996.

Becker, William H. "Ecological Sin." *Theology Today* 49 (July 1992): 152–64.

Berger, Peter L. *The Heretical Imperative: Contemporary Possibilities of Religious Affirmation.* Garden City, N.Y.: Anchor/Doubleday, 1979.

———. *The Sacred Canopy.* Garden City, N.Y.: Doubleday, 1969.

Berlin, Isaiah. *The Crooked Timber of Humanity: Chapters in the History of Ideas.* New York: Vintage Books, 1959.

Berman, Morris. *The Reenchantment of the World.* Toronto: Bantam Books, 1981.

Berry, Wendell. *Life Is a Miracle: An Essay against Modern Superstition.* Washington, D.C.: Counterpoint, 2000.

Blumenberg, Hans. *The Legitimacy of the Modern Age.* Translated by Robert M. Wallace. Cambridge: Massachusetts Institute of Technology Press, 1983.

Bordo, Susan. *The Flight to Objectivity: Essays on Cartesianism and Culture.* Albany: State University of New York Press, 1987.

Borg, Marcus J., and N. T. Wright. *The Meaning of Jesus: Two Versions.* San Francisco: HarperSanFrancisco, 1999.

Bowler, Peter J. *The Non-Darwinian Revolution: Reinterpreting a Historical Myth.* Baltimore: Johns Hopkins University Press, 1988.

Bronowski, Jacob. *The Origins of Knowledge and Imagination.* New Haven, Conn.: Yale University Press, 1978.

Brooke, John Hedley. *Science and Religion: Some Historical Perspectives.* Cambridge: Cambridge University Press, 1991.

Brown, Warren S. "Cognitive Contributions to Soul." Pages 99–125 in *Whatever Happened to the Soul?* Edited by Warren S. Brown, Nancey Murphy, and H. Newton Malony. Minneapolis: Fortress, 1998.

Browning, Donald. "The Challenge of the Future to the Science-Religion Dialogue." *Zygon: Journal of Religion and Science* 22 (Twentieth Anniversary Issue, 1987): 35–38.

Brueggemann, Walter. *Cadences of Home.* Louisville: Westminster John Knox, 1997.

———. *Finally Comes the Poet: Daring Speech for Proclamation.* Minneapolis: Fortress, 1989.

———. *Texts under Negotiation.* Minneapolis: Fortress, 1993.

———. *Theology of the Old Testament: Testimony, Dispute, Advocacy.* Minneapolis: Fortress, 1997.

Casey, Edward S. *Getting Back into Place: Toward a Renewed Understanding of the Place-World.* Bloomington: Indiana University Press, 1993.

Clapp, Rodney. *How Firm a Foundation: Can Evangelicals Be Nonfoundationalists?* Edited by Timothy R. Phillips and Dennis L. Okholm. Downers Grove, Ill.: InterVarsity Press, 1996.

Cohen, Bernard. *Revolution in Science.* Cambridge: Belknap Press of Harvard University Press, 1985.

Coleman, Richard J. *Issues of Theological Conflict.* Grand Rapids, Mich.: Eerdmans, 1972.

Cox, Harvey. *Many Mansions: A Christian Encounter with Other Faiths.* Boston: Beacon, 1988.

Crites, Stephen. "The Narrative Quality of Experience." Pages 65–88 in *Why Narrative?* Edited by Stanley Hauerwas and Gregory Jones. Grand Rapids, Mich.: Eerdmans, 1989.

D'Amico, Robert. *Historicism and Knowledge.* New York: Routledge, 1989.

Davis, Nuel Pharr. *Lawrence and Oppenheimer.* New York: DaCapo Press, 1986.

Dawkins, Richard. *Unweaving the Rainbow: Science, Delusion and the Appetite for Wonder.* Boston: Houghton Mifflin, 1998.

Dennett, Daniel C. *Darwin's Dangerous Idea: Evolution and the Meanings of Life.* New York: Simon & Schuster, 1995.

Desmond, Adrian. *Huxley: From Devil's Disciple to Evolution's High Priest.* Cambridge, Mass.: Harvard University Press, 1994.

Desmond, Adrian, and James Moore. *Darwin: The Life of a Tormented Evolutionist.* New York: W. W. Norton, 1991.

Deutsch, David. *The Fabric of Reality.* New York: Penguin, 1997.

Dewart, Leslie. *The Foundations of Belief.* New York: Herder & Herder, 1969.

———. *The Future of Belief: Theism in a World Come of Age.* New York: Herder & Herder, 1966.

Dobbs, Betty Jo Teeter, and Margaret C. Jacob. *Newton and the Culture of Newtonianism.* Atlantic Highlands, N.J.: Humanities Press International, 1995.

Dobzhansky, Theodosius. "Teilhard de Chardin and the Orientation of Evolution: A Critical Essay." Pages 229–48 in *Process Theology.* Edited by H. Cousins. New York: Newman Press, 1971.

Drees, Willem B. "Postmodernism and the Dialogue between Religion and Science." *Zygon: Journal of Religion and Science* 32, no. 4 (December 1997): 525–41.

———. "Naturalisms and Religion." *Zygon: Journal of Religion and Science* 32 (December 1997): 525–41.

———. *Religion, Science and Naturalism.* Cambridge: Cambridge University Press, 1996.

Dupré, Louis. *Religious Mystery and Rational Reflection.* Grand Rapids, Mich.: Eerdmans, 1998.

Dyson, Freeman J. *The Sun, the Genome, the Internet: Tools of Scientific Revolutions.* New York: Oxford University Press, 1999.

Eiseley, Loren. *The Firmament of Time.* New York: Atheneum, 1984.

Ellul, Jacques. *The Humiliation of the Word.* Translated by Joyce Main Hanks. Grand Rapids, Mich.: Eerdmans, 1985.

———. *Living Faith: Belief and Doubt in a Perilous World.* Translated by Peter Heinegg. San Francisco: Harper & Row, 1983.

———. *The Technological Bluff.* Translated by Geoffrey W. Bromiley. Grand Rapids, Mich.: Eerdmans, 1990.

Everdell, William R. *The First Moderns.* Chicago: University of Chicago Press, 1997.

Fackre, Gabriel. *The Doctrine of Revelation.* Grand Rapids, Mich.: Eerdmans, 1997.

———. "Narrative Theology: An Overview." *Interpretation* 37 (October 1983): 340–52.

Farley, Edward. *Deep Symbols: Their Postmodern Effacement and Reclamation.* Valley Forge, Pa.: Trinity Press International, 1996.

Farrell, Frank B. *Subjectivity, Realism and Postmodernism.* Cambridge: Cambridge University Press, 1996.

Ferris, Timothy. *Coming of Age in the Milky Way.* New York: William Morrow, 1988.

Foucault, Michel. "Nietzsche, Genealogy, History." Pages 76–100 in *The Foucault Reader.* Edited by Paul Rabinow. New York: Pantheon Books, 1984.

———. *The Order of Things: An Archaeology of the Human Sciences.* New York: Vintage/Random House, 1994.

Frei, Hans. *The Eclipse of Biblical Narrative: A Study in Eighteenth and Nineteenth Century Hermeneutics.* New Haven, Conn.: Yale University Press, 1974.

Frye, Northrop. *The Great Code: The Bible and Literature.* San Diego: Harcourt Brace Jovanovich, 1982.

———. *Words with Power: Being a Second Study of the Bible and Literature.* San Diego: Harcourt Brace Jovanovich, 1990.

Funkenstein, Amos. *Theology and the Scientific Imagination from the Middle Ages to the Seventeenth Century.* Princeton: Princeton University Press, 1988.

Galison, Peter. *Image and Logic.* Chicago: University of Chicago Press, 1997.

Geertz, Clifford. *Available Light: Anthropological Reflections on Philosophical Topics.* Princeton: Princeton University Press, 2000.

Gelpi, Donald L. *The Turn to Experience in Contemporary Theology.* Mahwah, N.J.: Paulist Press, 1994.

Gerhard, Mary, and Allan Russell. *Metaphoric Process.* Fort Worth: Texas Christian University Press, 1984.

Gilkey, Langdon. *Nature, Reality, and the Sacred: The Nexus of Science and Religion.* Minneapolis: Fortress, 1993.

———. "Religion and Science in an Advanced Scientific Culture." *Zygon: Journal of Religion and Science* 22 (June 1987): 165–78.

———. *Religion and the Scientific Future.* New York: Harper & Row, 1970.

———. "The Structure of Academic Revolution." Pages 538–46 in *The Nature of Scientific Discovery: Symposium Commemorating the 500th Anniversary of the Birth of Nicholaus Copernicus.* Edited by Owen Gingerich. Washington, D.C.: Smithsonian Institution Press, 1975.

Gillespie, Neal C. *Charles Darwin and the Problem of Creation.* Chicago: University of Chicago Press, 1979.

Gillispie, Charles Coulston. *The Edge of Objectivity.* Princeton, N.J.: Princeton University Press, 1960.

Gingerich, Owen. "Osiander's Introduction to *De Revolutionibus.*" Pages 301–4 in *The Nature of Scientific Discovery: Symposium Commemorating the 500th Anniversary of the Birth of Nicholaus Copernicus.* Washington, D.C.: Smithsonian Institution Press, 1975.

Gombrich, E. H. *Symbolic Images: Studies in the Art of the Renaissance.* Cambridge: Phaidon, 1972; New York: E. P. Dutton, 1978.

Gould, Stephen Jay. *Bully for Brontosaurus.* New York: W. W. Norton, 1991.

———. *Rock of Ages: Science and Religion in the Fullness of Life.* New York: Ballantine, 1999.

———. *Time's Arrow, Time's Cycle: Myth and Metaphor in the Discovery of Geological Time.* Cambridge, Mass.: Harvard University Press, 1987.

Grant, George Parkin. *Technology and Justice.* Notre Dame, Ind.: University of Notre Dame Press, 1986.

Green, Garrett. *Imagining God: Theology and the Religious Imagination.* San Francisco: Harper & Row, 1989.

Greene, John C. *Science, Ideology, and World View.* Berkeley: University of California Press, 1981.

Gregersen, Niels Henrik, and J. Wentzel van Huyssteen, eds. *Rethinking Theology and Science: Six Models for the Current Dialogue.* Grand Rapids, Mich.: Eerdmans, 1998.

Griffin, David Ray. *God, Power and Evil: A Process Theodicy.* Philadelphia: Westminster, 1971.

Gunton, Colin. *Enlightenment and Alienation.* Grand Rapids, Mich.: Eerdmans, 1985.

———. "All Scripture Is Inspired . . . ?" *Princeton Seminary Bulletin* 14 (1993): 240–53.

Hacking, Ian. *Emergence of Probability.* Cambridge: Cambridge University Press, 1965.

———. *Representing and Intervening.* Cambridge: Cambridge University Press, 1983.

Hardison, O. B., Jr. *Disappearing through the Skylight: Culture and Technology in the Twentieth Century.* New York: Viking Penguin, 1989.

Hartshorne, Charles, and Creighton Peden. *Whitehead's View of Reality.* New York: Pilgrim, 1981.

Harvey, Van H. *The Historian and the Believer.* New York: Macmillan, 1966.

Hauerwas, Stanley, and L. Gregory Jones, eds. *Why Narrative? Readings in Narrative Theology.* Grand Rapids, Mich.: Eerdmans, 1989.

Haught, John F. *The Cosmic Adventure: Science, Religion and the Quest for Purpose.* New York: Paulist Press, 1984.

———. *God after Darwin: A Theology of Evolution.* Boulder, Colo.: Westview, 2000.

———. *Science and Religion. From Conflict to Conversation.* New York: Paulist Press, 1995.

Hawking, Stephen. *A Brief History of Time: From the Big Bang to Black Holes*. Toronto: Bantam Books, 1988.

Hebblethwaite, Brian. "Providence and Divine Action." *Religious Studies* 14 (June 1978): 223–36.

Heim, Mark. *Salvation—Truth and Difference in Religion*. Maryknoll, N.Y.: Orbis, 1997.

Hensley, Jeffrey. "Are Postliberals Necessarily Antirealists?" Pages 69–80 in *The Nature of Confession*. Edited by Timothy R. Phillips and Dennis L. Okholm. Downers Grove, Ill.: InterVarsity Press, 1996.

Herrmann, Eberhard. "A Pragmatic Approach to Religion and Science." Pages 121–56 in *Rethinking Theology and Science: Six Models for the Current Dialogue*. Edited by Niels Henrik Gregersen and J. Wentzel van Huyssteen. Grand Rapids, Mich.: Eerdmans, 1998.

Hirschfelder, Joseph O. "The Scientific and Technological Miracle at Los Alamos." Pages 67–88 in *Reminiscences of Los Alamos, 1943–1945*. Edited by Lawrence Badash, Joseph Hirschfelder, and Herbert Broida. Dordrecht, Holland: D. Reidel, 1980.

Hodgson, Peter C. *God in History: Shapes of Freedom*. Nashville: Abingdon, 1989.

Holton, Gerald. *Einstein, History, and Other Passions: The Rebellion against Science at the End of the Twentieth Century*. Reading, Mass.: Addison-Wesley, 1996.

Howe, Ruel L. *The Miracle of Dialogue*. New York: Seabury, 1965.

Hull, David L. *Darwin and His Critics: The Reception of Darwin's Theory of Evolution by the Scientific Community*. Chicago: University of Chicago Press, 1973.

Hunsinger, George. *Disruptive Grace: Studies in the Theology of Karl Barth*. Grand Rapids, Mich.: Eerdmans, 2000.

Husserl, Edmund. *The Crisis of European Sciences and Transcendental Phenomenology*. Translated by David Carr. Evanston, Ill.: Northwestern University Press, 1970.

Jammer, Max. *Einstein and Religion*. Princeton: Princeton University Press, 1999.

Jenson, Robert W. *Systematic Theology*. 2 vols. Oxford: Oxford University Press, 1998–99.

Kaufman, Gordon D. *In Face of Mystery: A Constructive Theology*. Cambridge, Mass.: Harvard University Press, 1993.

Kaufmann, Stuart. *At Home in the Universe: The Search for Laws of Self-Organization and Complexity.* New York: Oxford University Press, 1995.

Keck, Leander E. *The Church Confident.* Nashville: Abingdon, 1993.

Kelsey, David H. "Whatever Happened to the Doctrine of Sin?" *Theology Today* 50 (July 1993): 169–78.

Kenseth, Arnold. *Sabbaths, Sacraments, and Seasons.* Amherst, Mass.: Windhover Press, 1982.

Kimel, Alvin F., Jr., ed. *Speaking the Christian God.* Grand Rapids, Mich.: Eerdmans, 1992.

Kliever, Lonnie D. *The Shattered Spectrum.* Louisville: John Knox Press, 1981.

Koestler, Arthur. *The Sleepwalkers.* New York: Macmillan, 1959.

Kohák, Erazim. *The Embers and the Stars.* Chicago: University of Chicago Press, 1984.

Krauthammer, Charles. "In Search of the Silver Bullet." *Time* (November 4, 1985): 98.

Kuhn, Thomas S. *The Copernican Revolution.* Cambridge: Cambridge University Press, 1957.

———. *The Essential Tension: Selected Studies in Scientific Tradition and Change.* Chicago: University of Chicago Press, 1977.

———. *The Structure of Scientific Revolutions.* 2d ed. Chicago: University of Chicago Press, 1970.

Küng, Hans, and David Tracy, eds. *Paradigm Change in Theology.* New York: Crossroad, 1989.

Lash, Christopher. *The True and Only Heaven: Progress and Its Critics.* New York: W. W. Norton, 1991.

Laudan, Larry. *Progress and Its Problems: Towards a Theory of Scientific Growth.* Berkeley: University of California Press, 1977.

Leiss, William. *The Domination of Nature.* Boston: Beacon, 1972.

Levenson, Jon D. *Creation and the Persistence of Evil: The Jewish Drama of Divine Omnipotence.* San Francisco: Harper & Row, 1969.

Lewis, C. S. *A Grief Observed.* New York: Bantam Books, 1976.

Lewontin, Richard. *The Triple Helix: Gene, Organism and Environment.* Cambridge, Mass.: Harvard University Press, 2000.

Lindbeck, George A. "Atonement and the Hermeneutics of Intratextual Social Embodiment." Pages 221–40 in *The Nature of Confession.*

Edited by Timothy R. Phillips and Dennis L. Okholm. Downers Grove, Ill.: InterVarsity Press, 1996.

———. *The Nature of Doctrine: Religion and Theology in a Postliberal Age*. Philadelphia: Westminster, 1984.

Lovejoy, Arthur O. *The Great Chain of Being: A Study of the History of an Idea*. Cambridge, Mass.: Harvard University Press, 1936.

Lyotard, Jean-François. *The Postmodern Condition: A Report on Knowledge*. Translated by Geoff Bennington and Brian Massumi. Minneapolis: University of Minnesota Press, 1984.

MacIntyre, Alasdair. *After Virtue: A Study in Moral Theory*. 2d ed. Notre Dame, Ind.: University of Notre Dame Press, 1984.

———. "Epistemological Crises, Dramatic Narrative, and the Philosophy of Science." Pages 152–57 in *Why Narrative?* Edited by Stanley Hauerwas and L. Gregory Jones. Grand Rapids, Mich.: Eerdmans, 1989.

———. "The Logical Status of Religious Belief." Pages 159–201 in *Metaphysical Beliefs*, edited by Stephen Toulmin. New York: Schocken Books, 1957.

———. *Whose Justice? Which Rationality?* Notre Dame, Ind.: University of Notre Dame Press, 1988.

Manuel, Frank E. *A Portrait of Isaac Newton*. Cambridge, Mass.: Belknap Press of Harvard University Press, 1968.

Martin, Calvin Luther. *In the Spirit of the Earth: Rethinking History and Time*. Baltimore: Johns Hopkins University Press, 1992.

Mayr, Ernst. *The Growth of Biological Thought*. Cambridge, Mass.: Belknap Press of Harvard University Press, 1982.

McFague, Sallie. "Models of God: Three Observations." *Religion and Intellectual Life* 5 (spring 1988): 9–44.

———. *The Body of God: An Ecological Theology*. Minneapolis: Fortress, 1993.

McKibben, Bill. *The End of Nature*. London: Penguin, 1990.

McMullin, Ernan. "Biology and the Theology of the Human Heart." Pages 367–93 in *Controlling Our Destinies: Historical, Philosophical, Ethical and Theological Perspectives on the Human Genome Project*. Edited by Phillip R. Sloan. Notre Dame, Ind.: University of Notre Dame Press, 2000.

Merchant, Carolyn. *The Death of Nature: Women, Ecology and the Scientific Revolution*. San Francisco: Harper & Row, 1983.

Midgley, Mary. *Science as Salvation.* New York: Routledge, 1992.

Migliore, Daniel L. *Faith Seeking Understanding.* Grand Rapids, Mich.: Eerdmans, 1991.

Moltmann, Jürgen. "The Alienation and Liberation of Nature." Pages 138–39 in *On Nature.* Edited by Leroy Rounder. Notre Dame, Ind.: University of Notre Dame Press, 1984.

———. *God in Creation: A New Theology of Creation and the Spirit of God.* San Francisco: Harper & Row, 1985.

Moore, Stephen D. *Poststructuralism and the New Testament.* Minneapolis: Fortress, 1994.

Morris, Richard. *Dismantling the Universe: The Nature of Scientific Discovery.* New York: Simon & Schuster, 1983.

Morrison, Roy D., II. *Science, Theology and the Transcendental Horizon: Einstein, Kant and Tillich.* Atlanta: Scholars Press, 1994.

Murphy, Nancey. *Anglo-American Postmodernity.* Boulder, Colo.: Westview, 1997.

———. *Beyond Liberalism and Fundamentalism: How Modern and Postmodern Philosophy Set the Theological Agenda.* Valley Forge, Pa.: Trinity Press International, 1996.

———. "Postmodern Apologetics: Or Why Theologians Must Pay Attention to Science." Pages 105–20 in *Religion and Science: History, Method, Dialogue.* Edited by Mark W. Richardson and Wesley J. Wildman. New York: Routledge, 1996.

———. *Theology in the Age of Scientific Reasoning.* Ithaca, N.Y.: Cornell University Press, 1990.

Nagel, Ernest. *The Structure of Science.* New York: Harcourt, Brace & World, 1961.

Newsom, Carol A. "The Moral Sense of Nature: Ethics in the Light of God's Speech to Job." *Princeton Seminary Bulletin* 15 (New Series 1994): 9–27.

Nygren, Anders. *Meaning and Method: Prolegomena to a Scientific Philosophy of Religion and a Scientific Theory.* Translated by Philip S. Watson. Philadelphia: Fortress, 1972.

Ormiston, Gayle L., and Raphael Sassower. *Narrative Experiments: The Discursive Authority of Science and Technology.* Minneapolis: University of Minnesota Press, 1989.

Ottati, Douglas F. *Hopeful Realism: Reclaiming the Poetry of Theology.* Cleveland: Pilgrim, 1999.

Pagels, Heinz R. *The Dreams of Reason: The Computer and the Rise of the Science of Complexity.* New York: Bantam, 1988.

Paley, William. *Natural Theology: or, Evidence of the Existence and Attributes of the Deity.* London, 1802.

Pannenberg, Wolfhart. *Anthropology in Theological Perspective.* Translated by Matthew O'Connell. Philadelphia: Westminster, 1985.

———. *Theology and the Philosophy of Science.* Philadelphia: Westminster, 1976.

———. *Toward a Theology of Nature: Essays on Science and Faith.* Louisville: Westminster John Knox, 1993.

Peacocke, Arthur. *Intimations of Reality: Critical Realism in Science and Religion.* Notre Dame, Ind.: University of Notre Dame Press, 1984.

———. "Relating Genetics to Theology on the Map of Scientific Knowledge." Pages 343–65 in *Controlling Our Destinies: Historical, Philosophical, Ethical and Theological Perspectives on the Human Genome Project.* Edited by Phillip R. Sloan. Notre Dame, Ind.: University of Notre Dame Press, 2000.

———, ed. *The Sciences and Theology in the Twentieth Century.* Notre Dame, Ind.: University of Notre Dame Press, 1981.

Peck, Scott. *The Road Less Traveled.* New York: Simon & Schuster, 1978.

Pera, Marcello. *The Discourses of Science.* Translated by Clarissa Botsford. Chicago: University of Chicago Press, 1994.

Peters, Ted. *God as Trinity.* Louisville: Westminster/John Knox, 1993.

———, ed. *Science and Theology: The New Consonance.* Boulder, Colo.: Westview, 1998.

Phillips, Timothy R., and Dennis L. Okholm, eds. *The Nature of Confession: Evangelicals and Postliberals in Conversation.* Downers Grove, Ill.: InterVarsity Press, 1996.

Placher, William C. *The Domestication of Transcendence: How Modern Thinking About God Went Wrong.* Louisville: Westminster John Knox, 1996.

———. *Unapologetic Theology.* Philadelphia: Westminster John Knox, 1989.

Polkinghorne, John. *Belief in God in an Age of Science.* New Haven, Conn.: Yale University Press, 1998.

———. “Creation and the Structure of the Physical World.” *Theology Today* 44 (April 1987): 53–68.

———. *Faith, Science and Understanding.* New Haven, Conn.: Yale University Press, 2000.

———. *Reason and Reality: The Relationship between Science and Theology.* Philadelphia: Trinity Press International, 1991.

———. *Science and Theology.* London: SPCK, 1998.

Polkinghorne, John, and Michael Welker, eds. *The End of the World and the Ends of God.* Harrisburg, Pa.: Trinity Press International, 2000.

Progoff, Ira. *At a Journal Workshop.* New York: Dialogue House Library, 1975.

Raymo, Chet. *Natural Prayers.* St. Paul: Hungry Mind Press, 1999.

Ricoeur, Paul. *Figuring the Sacred: Religion, Narrative and Imagination.* Minneapolis: Fortress, 1995.

———. *The Symbolism of Evil.* Boston: Beacon, 1967.

———. *Time and Narrative.* Translated by Kathleen McLaughlin and David Pellaue. 3 vols. Chicago: University of Chicago Press, 1984, 1985, 1988.

Rifkin, Jeremy. *Algeny.* New York: Viking, 1983.

———. *Entropy.* New York: Viking, 1980.

Roche, John. “Newton’s *Principia.*” Pages 43–61 in *Let Newton Be!* Edited by John Fauvel, Raymond Flood, Michael Shortland, and Robin Wilson. Oxford: Oxford University Press, 1988.

Rolston, Holmes, III. *Genes, Genesis and God: Values and Their Origins in Natural and Human History.* Cambridge: Cambridge University Press, 1999.

———. “Does Nature Need to Be Redeemed?” *Zygon: Journal of Religion and Science* 29 (June 1994): 205–29.

———. *Science and Religion.* Philadelphia: Temple University Press, 1987.

Rorty, Richard. *Philosophy and the Mirror of Nature.* Princeton: Princeton University Press, 1979.

Ruse, Michael. *The Darwinian Revolution.* Chicago: University of Chicago Press, 1979.

———. *Mystery of Mysteries: Is Evolution a Social Construction?* Cambridge, Mass.: Harvard University Press, 1999.

Santmire, H. Paul. *The Travail of Nature: The Ambiguous Ecological Promise of Christian Theology*. Philadelphia: Fortress, 1985.

Schama, Simon. *Dead Certainties: Unwarranted Speculations*. New York: Knopf, 1991.

Schmitz-Moormann, Karl. *Theology of Creation in an Evolutionary World*. Cleveland: Pilgrim, 1997.

Schumacher, E. F. *Small Is Beautiful: Economics as If People Mattered*. New York: Harper & Row, 1973.

Schweber, S. S. *In the Shadow of the Bomb*. Princeton: Princeton University Press, 2000.

Sheldrake, Rupert. *A New Science of Life*. Los Angeles: J. P. Tarcher, 1981.

Smart, James D. *The Interpretation of Scripture*. Philadelphia: Westminster, 1961.

Smith, Huston. *Beyond the Post-Modern Mind*. New York: Crossroad, 1982.

Smith, Wilfred Cantwell. *The Meaning and End of Religion: A New Approach to the Religious Traditions of Mankind*. New York: Macmillan, 1963.

Soskice, Janet Martin. "The Ends of Man and the Future of God." Pages 78–87 in *The End of the World and the Ends of God: Science and Theology on Eschatology*. Edited by John Polkinghorne and Michael Welker. Harrisburg, Pa.: Trinity International Press, 2000.

Stern, Robert. "MacIntyre and Historicism." Pages 146–60 in *After MacIntyre*. Edited by John Horton and Susan Mendus. Notre Dame, Ind.: University of Notre Dame Press, 1994.

Stout, Jeffrey. *The Flight from Authority*. Notre Dame, Ind.: University of Notre Dame Press, 1981.

Taylor, Charles. "Justice after Virtue." Pages 16–43 in *After MacIntyre*. Edited by John Horton and Susan Mendus. Notre Dame, Ind.: University of Notre Dame Press, 1994.

———. *Sources of the Self: The Making of Modern Identity*. Cambridge, Mass.: Harvard University Press, 1989.

Thiemann, Ronald F. *Revelation and Theology: The Gospel as Narrated Promise*. Notre Dame, Ind.: University of Notre Dame Press, 1985.

Torrance, Thomas F. *God and Rationality*. London: Oxford University Press, 1971.

Toulmin, Stephen. "The Historicization of Natural Science: Its Implication for Theology." Pages 233–41 in *Paradigm Change in Theology*, edited by Hans Küng and David Tracy. New York: Crossroad, 1989.

Toulmin, Stephen, and June Goodfield. *The Discovery of Time*. Chicago: University of Chicago Press, 1965.

———. *The Fabric of the Heavens: The Development of Astronomy and Dynamics*. New York: Harper & Row, 1961.

Tracy, David. *The Analogical Imagination: Christian Theology and the Culture of Pluralism*. New York: Crossroad, 1991.

———. *Blessed Rage for Order: The New Pluralism in Theology*. Minneapolis: Seabury, 1975.

———. *Plurality and Ambiguity: Hermeneutics, Religion, Hope*. San Francisco: Harper & Row, 1987.

Turner, James. *Without God, Without Creed: The Origins of Unbelief in America*. Baltimore: Johns Hopkins University Press, 1985.

van Huyssteen, J. Wentzel. *Duet or Duel? Theology and Science in a Postmodern World*. Harrisburg, Pa.: Trinity Press International, 1998.

———. *Essays in Postfoundationalist Theology*. Grand Rapids, Mich.: Eerdmans, 1997.

———. "Is the Postmodernist Always a Postfoundationalist?" *Theology Today* 50 (October 1993): 373–86.

———. *The Shaping of Rationality: Toward Interdisciplinarity in Theology and Science*. Grand Rapids, Mich.: Eerdmans, 1999.

Volf, Miroslav. "Theology, Meaning and Power: A Conversation with George Lindbeck on Theology and the Nature of Christian Difference." Pages 45–66 in *The Nature of Confession*. Edited by Timothy R. Phillips and Dennis L. Okholm. Downers Grove, Ill.: InterVarsity Press, 1996.

Wallace, Mark I. "The Wild Bird Who Heals: Recovering the Spirit in Nature." *Theology Today* 50 (April 1993): 13–28.

Ward, Graham. *Barth, Derrida and the Language of Theology*. Cambridge: Cambridge University Press, 1995.

Ward, Keith. *God, Chance and Necessity*. Oxford: Oxford University Press, 1996.

Weinberg, Steven. *The First Three Minutes: A Modern View of the Origin of the Universe*. New York: Basic Books, 1977.

Welker, Michael. "God's Eternity, God's Temporality, and Trinitarian Theology." *Theology Today* 55 (October 1998): 317–28.

Westermann, Claus. *Blessing in the Bible and the Life of the Church.* Philadelphia: Fortress, 1978.

Westfall, Richard S. *The Life of Isaac Newton.* Cambridge: Cambridge University Press, 1993.

———. *Science and Religion in Seventeenth-Century England.* New Haven, Conn.: Yale University Press, 1958.

Whitehead, Alfred. *Process and Reality.* New York: Macmillan, 1957.

Wilder, Amos N. "Story and Story-World." *Interpretation* 37 (October 1983): 360–63.

Wildiers, Max N. *The Theologian and His Universe.* New York: Seabury, 1982.

Willimon, William H. *Peculiar Speech: Preaching to the Baptized.* Grand Rapids, Mich.: Eerdmans, 1992.

Wink, Walter. *The Powers That Be.* New York: Galilee Doubleday, 1998.

Wolterstorff, Nicholas. *Divine Discourse: Philosophical Reflection on the Claim That God Speaks.* Cambridge: Cambridge University Press, 1995.

———. *Reason within the Bounds of Religion.* 2d ed. Grand Rapids, Mich.: Eerdmans, 1976.

———. *Until Justice and Peace Embrace.* Grand Rapids, Mich.: Eerdmans, 1983.

Zagorin, Perez. *Francis Bacon.* Princeton: Princeton University Press, 1998.

Index